Warren De la Rue

On the Total Solar Eclipse of July 18th, 1860,

observed at Rivabellosa, near Miranda de Ebro in Spain

Warren De la Rue

On the Total Solar Eclipse of July 18th, 1860,
observed at Rivabellosa, near Miranda de Ebro in Spain

ISBN/EAN: 9783337243845

Printed in Europe, USA, Canada, Australia, Japan

THE BAKERIAN LECTURE.

ON

THE TOTAL SOLAR ECLIPSE

OF JULY 18TH, 1860,

OBSERVED AT

RIVABELLOSA, NEAR MIRANDA DE EBRO,

IN SPAIN.

BY

WARREN DE LA RUE, Esq., Ph.D., F.R.S.,
Hon. Sec. Royal Astron. Soc., Treasurer Chem. Soc. &c.

From the PHILOSOPHICAL TRANSACTIONS.—Part I. 1862

LONDON:
PRINTED BY TAYLOR AND FRANCIS, RED LION COURT, FLEET STREET.
1862.

XVIII. The Bakerian Lecture.—*On the Total Solar Eclipse of July 18th, 1860, observed at Rivabellosa, near Miranda de Ebro, in Spain.*

By Warren De La Rue, *Esq., Ph.D., F.R.S., Hon. Sec. Royal Astron. Soc., Treasurer Chem. Soc., &c.*

Received January 30,—Read April 10, 1862.

My attention was first called to the Solar Eclipse of 1860, in the latter part of the year 1858, on the occasion of my visiting Russia, when Dr. Mädler placed in my hands a copy of his anticipative pamphlet, entitled "L'Eclipse Solaire du 18 Juillet, 1860."

This paper contained a Map of Spain, with certain lines indicating the position of the central path of the moon's shadow, the limits of totality, and its epoch at various localities; and it occurred to me, on perusing it, that, if circumstances should permit of my observing the eclipse, Santander would be very convenient for the disembarcation and erection of the instruments I should, in all probability, require for photographic observations, to the prosecution of which my successful researches in astronomical photography led me to think I ought to devote myself. On communicating my plans to Mr. Vignoles, he strongly recommended me to cross to the southern side of the Pyrenees in order to avoid the mists which are caused by the condensation of vapours from the ocean against the northern slopes of the mountains. Subsequently Mr. Vignoles published an eclipse-map of Spain on a very large scale, and I selected Miranda for my station; but he suggested that I should place my observatory at Rivabellosa, about two miles from that town.

It is fortunate that I changed my station from Santander to Rivabellosa, as many of those astronomers who selected the former place were prevented by the state of the atmosphere from observing the eclipse.

On my journey to Russia, I stopped at Königsberg and made the acquaintance of Dr. Luther, who showed me the Daguerreotype of the total eclipse of 1851, which had been taken by Dr. Busch with the Königsberg heliometer. Great credit is due to Dr. Busch for that successful pioneering experiment, more especially when due allowance is made for the uncertainty then existing as to the brilliancy of the prominences, and for the state of the photographic art at that epoch. In the interval of seven years, however, astronomical photography had made great progress; and I recollect being much struck with the very indifferent definition of the protuberances in the Daguerreotype, from which I inferred the impracticability of deriving any conclusive evidence respecting the nature of such appearances from photographs, unless more distinct ones could be obtained. The inspection of the Königsberg Daguerreotype subsequently exercised some influence on my plan of procedure. Discarding all thoughts of employing the Daguerreotype process, because the collodion process was far more sensitive and convenient, I chose the latter as best suited to my purpose, although I knew perfectly well, from experience, how frequently the collodion-film is rendered defective by specks, streaks, and even minute holes. It was open to me to employ an achromatic or a reflecting telescope of ordinary construction, and to place the sensitive plate in the principal focus; but I was aware that the largest telescope I could possibly take with me would only give an image of a very moderate size, and that any of the before-named defects in the collodion might fall over and obliterate, or so confuse the impression of any prominence in one photograph, as to render its identification with its impression in a subsequent photograph a matter of impossibility. These considerations led me to think that it would be very desirable to employ the Kew photo-heliograph, because in this instrument the primary focal image of the sun is enlarged from about half an inch in diameter to nearly 4 inches, which is a scale amply sufficient to counterbalance the disadvantages of the collodion process; but, on the other hand, the light is thus attenuated sixty-four times, besides being absorbed to some extent in passing through the two lenses composing the secondary magnifier, an ordinary Huyghenian eyepiece; and this question consequently presented itself, Would it be possible with such an enfeebled image to get even a single impression during the whole duration of the totality? This was an extremely doubtful matter. By employing the Kew heliograph one would evidently run the risk of returning without any pictures of the *totality*, however many might be procured of the other phases of the eclipse *.

At the meetings of the Astronomical Society, and on other occasions, I made inquiries of those astronomers who had witnessed the eclipse of 1851, respecting the intensity of the light of the corona and red flames, as compared with that of the moon, and the relative brightness of the one to the other; but their answers did not tend to increase my hopes in respect of the possibility of procuring photographs of the totality by means of the Kew instrument. The general impression I formed from the information thus derived was, that the light emitted by the corona and red flames, taken together, was about equal to that of a full moon—less rather than greater; but no one recollected precisely the brightness of the prominences as compared with that of the corona. With this imperfect information as a guide, an attempt was made at Kew to photograph the moon, but not the slightest impression could be procured of our satellite by an exposure of the sensitive plate, during one whole minute†, to its image in the heliograph. My expectation of success in getting pictures of the totality was not great after this trial; nevertheless I still thought it desirable to carry on the experiment to the end, on account of the value of the results if I should fortunately succeed. It occurred to me several times to fit up also a photographic apparatus to an achromatic telescope, but I finally concluded that to attempt too many things would be sure to result in complete failure. I endeavoured, however, to stimulate other astronomers to

* Report on Celestial Photography, by the author, in the Reports of British Association, 1859, p. 152.

† While this paper was passing through the press a very faint impression of the moon was procured with the Kew heliograph in three minutes, with chemicals which gave a very strong impression of it in four seconds in the focus of my reflector of 13 inches aperture and 10 feet focal length.—August 1862.

make photographic experiments in the manner which I have indicated, as offering a greater probability of at least a partial success, so that the chances of obtaining pictures might be multiplied. With this object, I circulated as rapidly as I could my Report to the British Association on "Celestial Photography," which passed through the press in May 1860, and of which copies were extensively sent both to English and foreign astronomers at the latter end of May and the beginning of June.

I have now in this narrative to go back some months earlier than the period just alluded to, in order to connect it with the Himalaya Expedition, an expedition originating solely with, and organized by, the Astronomer Royal. When the year 1859 was drawing to a close, and I was turning my thoughts to the preparation which would be required for July 1860, Mr. Airy mentioned that, if I had any intention of observing the eclipse, he might possibly be in a position to afford me some facilities for so doing, as he had it in contemplation to make an application to the Admiralty for a ship to convey intending observers to Spain, in the event of a sufficient number of astronomers expressing their willingness to join the expedition which he intended to organize. I expressed the satisfaction I felt in learning that he, the official head of astronomy in England, was willing to take the matter in hand, because I felt persuaded that, under his general direction, the expedition would prove a successful one; and I at once volunteered to form one of his party.

It was intimated to me that, in the event of my taking charge of the Kew heliograph, I should not be expected to entail upon myself the expenses of fitting it up for the object contemplated, or the personal expenses of the assistants who might accompany it, it having been from the first intended, that a grant from the Government Fund should be asked for to defray these charges. When, therefore, I had finally decided on taking charge of the instrument, I was requested to propose such a sum as I thought fully adequate to the purpose, and I named £150, which was granted. The entire expenses of the photographic expedition amounted to more than three times that sum, the balance being defrayed by myself.

The actual preparations were commenced at the latter end of January 1860, first of all by Mr. Beckley, the mechanical assistant of the Kew Observatory, and were continued, so far as his other occupations would permit, until the month of June; but, in spite of every exertion on his part, so much remained to be done, that in June I engaged Mr. Reynolds (now my private photographic assistant) to aid in completing the arrangements. My party, besides myself, was, after a few changes, thus finally constituted:—Mr. Beckley, Mr. Reynolds, Mr. Downes, and Mr. E. Beck.

Among the preparations to be made was a stand for the telescope, the cast-iron pedestal of the Kew heliograph being too heavy for convenient transport. It was necessary, moreover, to make some contrivance for supporting the frame of the polar axis in a position adapted to certain limits of latitude, within which I might fix my station; and it was thought that this could be best arranged by making a new cast-iron pedestal composed of several pieces which took apart for the convenience of carriage*.

* This iron stand has been left *in situ*, and thus marks the precise locality of my observatory.

Originally, merely a temporary tent in which to develope the photographs was procured; but when it was known that H.M.S. 'Himalaya' would be placed at the disposal of the Astronomer Royal, I put this aside, and caused a complete photographic observatory to be constructed, part to contain the heliograph with a removable roof, and part divided off and fitted up as a photographic room, with a cistern, to be filled from the outside, a sink, and with tables and shelves to hold the apparatus and photographs. This observatory took to pieces, and every part was marked when in its place, so that no time need be lost in putting it together again in its destined position. Besides the ordinary roof, there was another covering, consisting of strong canvas, supported at the distance of about three feet from the walls and roof of the developing-room. The object of this was to prevent the overheating of the photographic room, a circumstance most detrimental to photography. This canvas was kept wetted with water, in order that the evaporation might lower the temperature of the stratum of air between it and the observatory, and it fulfilled the object perfectly. The canvas, when the observatory was not in use, was drawn over the room containing the heliograph, and protected the instrument from rain.

The print exhibits the arrangement of the observatory, when secured after the day's work. The position of the canvas, and the simple arrangement for maintaining it at the proper distance from the house, and also the outside cistern, are well shown in the picture, which is copied from a photograph taken on the occasion of Mr. AIRY's visit to my station. When the observatory was at work, the canvas was removed from the front (the south side) and tied back as far as the upright which is seen on the western side, the top boards also being removed. The front boards were of a height which admitted of our observing the sun above them whenever it was desirable to do so.

Photographic chemicals were prepared in duplicate; part of the collodion intended to be used was mixed with the iodizing solution in London, and after subsidence was carefully decanted previous to packing, in order to avoid the defects before alluded to; but collodion and iodizing solution were also taken separately, so that some might be prepared on the spot, and used, if found free from defects, in that state of extreme sensitiveness which exists in collodion freshly iodized with the potassium iodizer. Nitrate-of-silver baths, prepared in the ordinary way with crystallized nitrate of silver, were taken, and were used in depicting the several phases of the eclipse, with the exception of those of totality. In taking the latter pictures the baths used were made with nitrate of silver which had been fused carefully in my own laboratory, and were so extremely sensitive that they would give photographs of the full moon in the focus of my reflector in less than a second of time, while with the usual bath five seconds were barely sufficient to give a picture of similar intensity.

As few astronomers perhaps are aware of the number of materials required for such an expedition, I here give the list of contents of one of the boxes of chemicals.

Packages.	Contents.	Packages.	Contents.
3	Half-pint bottles of Collodion.	1	½ oz. Iodide of Potassium.
1	Four-ounce Bottle of Collodion, iodized.	1	Ounce Measure.
1	Half-ounce Bottle of Pyrogallic Acid.	1	Gallon Distilled Water.
1	Six-ounce Bottle of Acetic Acid.	1	Set of Scales and Weights.
1	1½-pound Bottle of Hyposulphite of Soda.	3	Plate-drainers.
1	Case containing Oxide of Silver and dilute Nitric Acid, in separate bottles, for correcting the bath, in case of need.	1	4 oz. of Tripoli.
2	24 oz. of Nitrate-of-silver Bath.	1	Packet Cotton Wool.
1	2 oz. Crystals Nitrate of Silver.	1	Glass Funnel.
1	4 oz. fused Nitrate of Silver.	1	Retort Stand.
		1	Lantern.
		3	Bottles of Varnish.
			Test Papers.
			Filtering Paper.
		4	8-oz. Mixing Glasses for Collodion.

The apparatus, when completed, weighed 34 cwt., and was made up into thirty packages for convenience of transport. Among the miscellaneous requisites were included:—distilled water weighing 139 lbs.; engineers' and carpenters' tools weighing 113 lbs.; lanterns, lamp-oil, spirit-lamp, and spirits of wine, weighing together 73 lbs.; a small stove and kettle for boiling water, and, lastly, some preserved provisions, in case the party should be compelled to encamp for a few days. Owing, however, to the excellent arrangements most kindly made by Mr. VIGNOLES, the latter were quite unnecessary.

As my plans became matured, it occurred to me that it would be desirable to make determinations of geographical position; and I therefore borrowed from the Kew Observatory a small transit theodolite with 6-inch altitude and azimuth circles, both reading with the verniers only to one minute of arc. The optical part of the instrument was found to be very indifferent, and the readings of the altitude circle were, from some cause, not so accordant from time to time as they ought to have been. I took with me also three chronometers—a box chronometer, a pocket chronometer, both indicating Greenwich mean time, and my journeyman sidereal chronometer.

I was induced to make preparations for eye-observations (which I did not originally contemplate), partly in order that I might be in a position to interpret from my own sketches and recollections the results of the photographs, and partly because in case I should fail in making photographs, I might still be able to contribute something to the series of optical observations. I therefore took with me a beautiful achromatic of 3 inches' aperture, which Mr. DALLMEYER had kindly lent me for the occasion. This telescope was mounted by Messrs. TROUGHTON and SIMMS on a most convenient and steady altazimuth-stand, designed by the Astronomer Royal. The equatorial movement was effected by the joint action of two radius bars, which enabled me to keep the sun exactly within a tangential square, which I had had ruled upon glass and placed in the focus of an eyepiece to be described hereafter.

Lastly, through the kindness of Messrs. ELLIOTT and Mr. CASELLA, I obtained the loan of some meteorological instruments. Messrs. ELLIOTT lent me one of their excellent aneroid barometers. Mr. CASELLA lent me a marine barometer and a standard thermometer, both verified at Kew, the readings of which were used in the reductions of the astronomical observations.

The apparatus was sent to Plymouth on the 5th of July, whence we set sail on the 7th; on the 9th we reached our destination. Mr. VIGNOLES met the 'Himalaya' in a small steamer which he had chartered to convey ashore the astronomers who intended to land at Bilbao, with their apparatus and luggage; but I am placed under a further obligation to him, not only for his kind and liberal hospitality during my stay at Bilbao, but also for dispatching my apparatus, as soon as it was landed, to Rivabellosa, which is situated at a distance of seventy miles from the port of Bilbao, and is only accessible through a pass difficult for the transmission of heavy baggage.

On the evening of the 10th we left Bilbao in a diligence which I had engaged to convey my party to Rivabellosa, at which place we arrived on the next day, after a journey very trying to our chronometers.

THE VILLAGE OF RIVABELLOSA.

Printed by the ordinary Letterpress, from a block produced by means of Photography and Electrometallurgy
absolutely untouched by the graver.

The instruments reached Rivabellosa on the evening of the 11th; the previous part of the day had been occupied in taking a general survey of the country around the village, with the object of selecting a site whereon to erect our observatory, and I at length settled on one of the thrashing-floors* which are to be seen in great numbers in that country in the open fields. It was about 60 feet in diameter, and close to the road, which we found to be a great convenience, inasmuch as the water required for our use had to be brought from a distance. Moreover, it was level, and extremely hard and dry. I had hardly selected this site when I learned, with some concern, that the harvest had commenced, and that the proprietor intended to make use of the floor on the morrow for his thrashing operations, which it is the custom of the country to complete immediately after the reaping. However, Don SIMON, land-surveyor of the Bilbao and Tudela Railway, who explained this to me, kindly undertook to negotiate for the hire of the station: but the owner, when informed that his thrashing-floor was the best adapted for my purpose of any place I had seen, at once said that it was quite at my disposal, and, although he had to convey his grain to a distance, refused any remuneration.

The instruments were conveyed to the thrashing-floor, and the transit theodolite unpacked in time to make an observation of the sun soon after 10 o'clock on the morning of the 12th. By the evening the observatory was erected and in actual operation, and a photograph of the sun was obtained on the 14th. To my staff was at last added Mr. S. CLARK, who had acted as interpreter, and who kindly volunteered his assistance during the eclipse. And I am greatly indebted to the gentlemen composing my staff for their most efficient assistance. With a self-denial hardly to be expected under the circumstances, each carried out steadily his allotted task, and thus contributed to a result which is most gratifying, after so much preparation and trouble.

Upwards of forty photographs were taken during the eclipse, and a little before and after it,—two being taken during the totality, on which are depicted the luminous prominences, with a precision as to contour and position impossible of attainment by eye-observations.

Photographs of the sun were taken on the two days succeeding the eclipse, namely, the 19th and 20th, and the instruments were then taken down and packed.

On the 26th the Himalaya Expedition re-embarked on board the 'Himalaya,'—myself, staff, and baggage being reconveyed to the vessel from Bilbao through the kindness of Mr. VIGNOLES, who accompanied us to England. On the 28th we landed at Portsmouth. I must here take occasion to express my best thanks to Captain SECCOMBE and the officers of the 'Himalaya' for their great kindness during the outward and homeward voyage. My thanks are also due to Señor MONTESINO, Chairman of the Bilbao and Tudela Railway, for assistance rendered; also to Don SIMON, and to Mr. BENNISON and Mr. PRESTON, gentlemen belonging to Mr. VIGNOLES's staff.

* In the accompanying print a thrashing-floor, similar to the one I employed, is represented. The Plate is inserted partly with the object of calling attention to Mr. PAUL PRETSCH's phototype process, which I have recently employed to furnish representations of sun-spots. (See Monthly Notices Roy. Ast. Soc. vol. xii. p. 278.)

After having obtained the photographs and got them home safely, my work was only begun. I knew that they contained in themselves most valuable records; but it did not, in the first instance, appear so clearly how I could turn them to account. The two totality pictures presented the most interest, and to them I first turned my attention; and as it was evident that no measurements ought to be made on the originals, I then bethought me of the best means of multiplying them. In the first instance, I got some enlarged positive copies photographed by Mr. Downes, and, made some little progress in measuring them; but I soon found that I should require others, and, on attempting to have some made a little later in the year, I experienced an amount of difficulty I never could have anticipated. The original negatives proved to be so extremely intense, that nothing short of unobscured sunlight would penetrate them and reveal their details; so that it was only by working for many selected days with Mr. Downes, that I succeeded in getting a sufficient number of positive copies for my purpose. Those who remember how remarkably dull and wet the year 1860 was in England will readily understand that the selected days were few and far between, so that it was fast approaching winter before I had got on much with the work on the photographs. I also had recourse to the albumen process, and obtained a few copies of the size of the original by superposition, without the intervention of the camera. These were made in my presence, at Messrs. Negretti and Zambra's, and from these positives some negatives were taken. Although these copies did not aid me greatly, it was fortunate they were taken; for in course of time the original negative No. 26, the second of totality, gave indications of decay, and on attempting to save it by revarnishing it, the collodion expanded, and crinkled up so much that, except as a record of what was done, the original negative is spoiled. I have protected it from further injury by cementing it with Canada balsam to a second glass plate: but one of the albumen negatives must now supplement it, if more copies are ever taken by direct superposition. Enlarged negatives, however, exist; but, as something is always lost in copying, the damage to the original negative is unfortunate*.

Measurements of the enlarged positives on glass of the totality pictures soon proved to me that the most accurate results could be obtained by measuring the photographs of the other phases, and that these results would indicate the path of the moon, and the position of the centres at the epoch of totality, independently of any determinations of geographical position. For this object it became necessary to measure the original negatives, because the slightest deviation of the optical axis of the copying camera from a right angle to the plane of the sensitive plate, or the least eccentricity, would cause a distortion of the cusps. I had, therefore, to devise an instrument for measuring the photographs; and having considered how the object could be effected, I put one in hand with Messrs. Troughton and Simms, who constructed it for me with their usual skill and precision. After the instrument had been made, it was found to be convenient that some of the parts should be provided with a means of adjustment; so that although it was commenced in February of 1861, it was not until July 18th that it was

* The whole of the original negatives have been deposited with the Astronomer Royal.

completely finished and ready for use. Since that time, every spare moment has been devoted to the final accomplishment of this work; and, taking into account the interruptions I am subject to, I feel convinced that it could not have been done in less time, although I candidly confess that the delay in sending in this Report must appear scarcely warranted.

Determination of the Geographical Position.

Previous to starting for Spain, I made certain preparations for facilitating the after-operations which I might have to carry out; for I knew that the time allowed for getting the instruments into position would not be more than sufficient, even if each day permitted of observations being made. For this reason, I computed, with the formula

$$\frac{\sin(\alpha \mp \lambda)}{\sin \alpha} \quad (\alpha = \text{the polar distance, } \lambda = \text{the colatitude}),$$

a series of star constants for all the stars likely to be visible in my instrument. I also computed a table of corrections, to be applied to the times of equal altitudes of the sun for intervals of two, three, four, five, and six hours, for each day from July 12 to July 22; and similar corrections to the azimuths of equal altitudes, to enable me to lay off at once a meridian line and erect a mark. My star constants did not, however, prove of much service; for it was rarely that I could get a glimpse of the stars; so that on only two occasions was I enabled to make any observations at all, and then only with the greatest difficulty, although I watched patiently for opportunities through or between the clouds.

I have already stated that I took with me three chronometers. My journeyman sidereal chronometer, Leplastrier 2915, is a very old one. In consequence of the wear of the fusee, this chronometer, which is an eight-day one, varies in its daily rate from 8·6 seconds to 17 seconds, losing; but during long periods the rate is pretty uniform, as will be seen from the following observations:—

Leplastrier 2915.

	sec.
From June 30 to July 3, losing daily	10·36
From July 3 to July 30, losing daily	12·51
From July 30 to August 7, losing daily	12·06
From August 7 to August 17, losing daily	11·29
On July 3, 21^h 24^m, it was fast of Cranford sidereal time	194·8
Cranford observatory is west of Greenwich in longitude	97·5
Hence, on July 3, 21^h 24^m, Leplastrier No. 2915 was fast of Greenwich sidereal time	97·3
	194·8
On July 30, 6^h 30^m, it was slow of Cranford sidereal time	138·0

The pocket chronometer, Frodsham No. 9768, was found to have the following rates and errors:—

	sec.
From June 26 to June 29, its daily rate was losing	1·05
From June 29 to June 30, its daily rate was losing	2·84
From June 30 to July 3, its daily rate was losing	2·00
On June 26, 15h, it was slow of Greenwich mean time	3·0
On July 3, 15, it was slow of Greenwich mean time	15·0

From these data, the mean daily rate was 1·71 seconds losing. On the 17th or 18th of July this chronometer tripped, most likely in consequence of its having been touched by the inmates of my lodgings; and it was therefore useless to make any determination of its error on my return from Spain.

From a comparison with the box chronometer to be next mentioned, for the loan of which I was indebted to the kindness of Mr. Frodsham, I estimated the error on Greenwich mean time of Frodsham 9768 to be 3·2 sec. slow, for July 3, 15h, instead of 3 sec., as found by observation. The comparison was made on June 25, 22h: two hours later Mr. Ellis found the box chronometer, Frodsham

	sec.	
No. 3094, to be on June 26, 0h	1·9	slow of Greenwich mean time.
In two hours 3094 would have gained	0·08	
At the time of comparison 9768 was slow of 3094	0·50	
Between the comparison and the ascertaining of the error of 9768, namely 17 hours, the latter would lose	0·74	
Estimated error of Frodsham 9768, on June 26, 15h	3·22	

This difference between 3·2 sec. and 3·0 sec. might have been occasioned by the journey of No. 3094 to Greenwich, but in any case was so small that, for the subsequent calculations, I retained without correction the error afterwards determined by observation on July 3, 15h, when Frodsham 9768 was slow of Greenwich mean time 15·0 sec.; whence on July 5, 0h, its error would have been 17·3 sec. slow of Greenwich mean time.

The box chronometer, Frodsham No. 3094, was (through the kindness of the Astrononomer Royal) compared at Greenwich by Mr. Ellis, whose determination of its errors are given below.

	h	m	sec.	
June 26	0	0	1·9	slow of Greenwich mean time.
June 27	0	0	1·0	
June 28	0	0	0·5	
June 29	0	0	1·3	fast of Greenwich mean time.
June 30	0	0	2·3	
July 1	0	0		
July 2	0	0	3·6	
July 3	0	0	4·5	

The average daily rate of this chronometer was therefore gaining 0·91 sec.; and assuming this rate to have continued, on July 5 0^h its error would have been fast of Greenwich mean solar time 6·3 seconds.

On the supposition that the pocket chronometer would continue to lose 1·71 sec. daily, and that the box mean-time chronometer would continue to gain daily 0·91 sec.

box chronometer Frodsham 3094 would gain over Frodsham 9768 . . 2·62 secs. daily.

The subjoined Table contains the assumed errors of each chronometer, the estimated difference between the two chronometers, and the difference actually observed as nearly as possible at noon of each day—the observation being reduced to noon:—

Date.	Assumed error of Frodsham 3094.	Assumed error of Frodsham 9768.	Estimated difference between 9768 and 3094.	Observed difference between 9768 and 3094.
	sec.	sec.	sec.	sec.
July 5.	fast 6·3	slow 17·3	−23·6	−24·5
July 6.	fast 7·2	slow 19·0	−26·2	
July 7.	fast 8·1	slow 20·7	−28·8	
July 8.	fast 9·0	slow 22·4	−31·4	
July 9.	fast 9·9	slow 24·1	−34·0	−35·0
July 10.	fast 10·8	slow 25·8	−36·6	−37·0
July 11.	fast 11·7	slow 27·5	−39·2	−37·5
July 12.	fast 12·6	slow 29·2	−41·8	−37·5
July 13.	fast 13·6	slow 30·9	−44·5	−38·5
July 14.	fast 14·5	slow 32·6	−47·1	−40·0
July 15.	fast 15·4	slow 34·3	−49·7	−41·5
July 16.	fast 16·3	slow 36·1	−52·4	−42·5

On examining the two box chronometers immediately after our arrival at Rivabellosa, Leplastrier 2915 was found to be apparently uninjured, but I was chagrined to find that Frodsham No. 3094 had been most severely disturbed by the joltings of our vehicle, notwithstanding the protection of its outside padded case, and an extra precaution I had taken to press shavings into its own case, to keep it firm in its place. The cap of the glass had become unscrewed, the glass had shaken out, and the chronometer itself, shifting from its normal position, had risen out of its seat; fortunately, however, the glass could not move far, on account of the wadding, and the hands were consequently uninjured. I succeeded in replacing the chronometer, and in putting the glass into its frame; but it thenceforward took up an entirely new rate, as was evident on comparing the differences between its readings and those of the pocket chronometer. An inspection of the foregoing Table shows that up to the 10th the two chronometers maintained the average rate assigned to each; for example, the computed difference minus the observed difference on that day amounted to only −0·4 second. After my return to England, chronometer No. 3094 was, with the Astronomer Royal's kind permission, again compared by Mr. Ellis, who found the following errors from Greenwich mean time:—

For Frodsham 3094.

	h	m	secs.
August 10	0	0	2·8 slow.
August 11	0	0	3 slow.
August 13	0	0	2·7 slow.
August 14	0	0	3 slow.
August 15	0	0	2·2 slow.
August 16	0	0	2·6 slow.
August 17	0	0	2·8 slow.
August 18	0	0	2·7 slow.
August 20	0	0	2·5 slow.

its average daily rate being losing 0·03 second, whereas, before starting, it was gaining 0·91 second.

On the 16th of July, Mr. OTTO STRUVE, Dr. WINNECKE, and Lieut. OOM visited my station; and I took advantage of their doing so, to make a comparison of chronometers. Dr. WINNECKE estimated the error of Frodsham No. 3094 to be fast of Greenwich mean time 7·5 seconds; consequently as the pocket chronometer Frodsham No. 9768 was slow of No. 3094 42·5 seconds, it follows, from this comparison, that it was −42·5 seconds +7·5 seconds =35 seconds slow of Greenwich mean time, which differs by only −1·1 second from the error assigned to it, by applying its mean daily losing rate of 1·71 second.

It appears, therefore, that the pocket chronometer could be relied on up to the 16th of July. On correcting the assumed error from Greenwich mean time of No. 3094 by comparisons with the pocket chronometer, we obtain the following:—

Frodsham No. 3094.

	Computed error from Greenwich mean time, taking No. 9768 as the standard.	Assumed error from Greenwich mean time applying its average rate.
July 9	fast 10·9 seconds	fast 9·9 seconds
July 10	fast 11·2 seconds	fast 10·8 seconds
July 11	fast 10·0 seconds	
July 12	fast 8·3 seconds	
July 13	fast 7·6 seconds	
July 14	fast 7·4 seconds	
July 15	fast 7·2 seconds	
July 16	fast 6·4 seconds	

With the advantage of the comparison made on the occasion of Mr. STRUVE's visit, I have confidence in assuming the error of Frodsham 3094 to have been from +6·4 to +7·5 seconds, say +7 seconds, on July 16 at noon.

On the 19th the Astronomer Royal and his party honoured me with a visit, and I had the great satisfaction of hearing from our leader that he was well satisfied with the success of my photographic operations, and also with the arrangements of the observatory, and the many preparations which had been made to secure the result.

At 2^h 30^m, a comparison was made between Mr. AIRY's pocket chronometer Molyneux No. 1007 and the box chronometer Frodsham No. 3094; Molyneux was slow of Frodsham 3094 43 seconds.

Molyneux 1007 was slow of Greenwich mean time,

	h	sec.	
On July 16 .	23	28·8	losing daily 7·2 seconds.
July 17 . .	22	35·7	gaining daily 2·7 seconds.
July 18 . .	22	33·0	

Applying the latter rate, Molyneux would appear to have been, on July 19, 2^h 30^m, 32·5 seconds slow, and consequently Frodsham 3094 10·5 seconds fast of Greenwich mean time. I believe that Molyneux 1007 could not be greatly depended on; but, the comparison of chronometers having been made, I place the result on record, although I am not able to make it accord with the other observations within several seconds.

Observations.

The following observations were made with the transit theodolite. During the day the instrument had to remain exposed to the sun; and this caused the several parts to expand very unequally, and kept the bubble in the level always in motion—a circumstance which proved very troublesome.

Estimation of Longitude.

July 12. Four pairs of reduced observations of equal altitudes of the sun showed that at local mean noon

Frodsham 3094 was fast of local mean time . . . 11 min. 51·9 sec.

July 14. Two pairs of reduced observations of equal altitudes of the sun showed that at local mean noon

Frodsham 3094 was fast of local mean time . . . 11 min. 51·3 sec.

With reduced observations of the azimuths of equal altitudes of the sun on the 12th and the 14th, northern and southern adjustable meridian marks were placed, the first against a building, the second against some trees; both sufficiently distant to give distinct vision of the mark, which was a cross × of wood, moveable in a top and a bottom groove in a wooden frame.

Attempts were made on the night of the 13th to obtain observations of stars; but the weather was too cloudy; by dint of perseverance, however, I did manage to get, through breaks in the clouds, the meridian altitude of α Lyræ, and an altitude of the pole star out of the meridian, presently to be referred to in the determinations of latitude.

On the night of the 14th I was more fortunate, and was able to obtain observations of a high and low star, and finally to adjust the meridian marks by means of an observation of δ Ursæ Minoris on the meridian. As soon as the observation of δ Ursæ Minoris was made, an assistant at each of the marks held a lantern near it, so that being illuminated I might see it with the theodolite and thus be enabled to make signals to them for the permanent adjustment of the marks, which was successfully accomplished. The following day being Sunday, no work was done; but on Monday Mr. PRESTON volunteered to aid me in projecting my meridian towards the station of the Astronomer Royal at Pobes. Using, as a signal flag, a bed sheet, which was not at all larger than was necessary in order to be well seen, I was able to direct Mr. PRESTON where to erect a staff with a similar appendage, in the line of my meridian, towards the north, and at a distance of about seven miles from Rivabellosa. Mr. O. STRUVE, as I before said, visited us on the 16th and undertook to work out the geodetic survey, which it was agreed should be made to connect the Pobes and Rivabellosa stations.

July 14. Observations of α Scorpii and ζ Herculis, when reduced, showed that at 16 hour 28 min.

Leplastrier 2915 was fast of local sidereal time . . . . . 11 min. 1·7 sec.

The weather afforded no other opportunity for star observations for longitude determinations until the night of the 18th, when I was too much fatigued to avail myself of it.

July 16, 0 h. A transit of the sun showed that

Frodsham 3094 was fast of local mean time . . . . . . 11 min. 49·9 sec.

July 20, 0 h. A transit of the sun showed that

Frodsham 3094 was fast of local mean time . . . . . . 11 min. 41·5 sec.

July 20, 8 h. 0 min. sidereal time. The transit of the sun being observed simultaneously with the sidereal chronometer, showed that

Leplastrier 2915 was fast of local sidereal time . . . . 9 min. 55·8 sec.

From the foregoing observations are derived the following results.

Frodsham No. 3094.

	Fast of Rivabellosa, mean solar time.			Daily rate.	Error on Greenwich mean time as estimated by comparison with Frodsham 9768.
	h	m	sec.	sec.	sec.
July 12th . .	0	11	51·9		Fast 8·3
				Losing 0·3	
July 14th . .	0	11	51·3		Fast 7·4
				Losing 0·7	
July 16th . .	0	11	49·9		Fast 6·4
				Losing 2·1	
July 20th . .	0	11	41·5		

whence the Longitude was West of Greenwich

	m	sec.
July 12 . . .	11	43·6
July 14 . . .	11	43·9
July 16 . . .	11	43·5
Mean . . .	11	43·7

Taking Dr. WINNECKE's estimate of the error of Frodsham No. 3094 on the 16th, namely, 7·5 seconds fast of Greenwich, we get for the longitude

West 11 min. 42·4 seconds

Applying the average daily losing rate of 12·51 sec. since July 3rd, 21 h. 24 min. for Leplastrier, we derive the following results from the observations made with that chronometer :—

Leplastrier No. 2915

	h	m	Was fast of local sidereal time. m	sec.	Error on Greenwich sidereal time by applying mean daily rate.	m	sec.
July 14 . .	16	47	11	1·7	Slow	0	37·9
July 20 . .	8	0	9	55·8	Slow	1	48·4

whence the longitude was West of Greenwich

	m	sec.
July 14 . .	11	39·6
July 20 . .	11	44·2
Mean .	11	41·9

By combining all the foregoing results, and taking the arithmetical mean, we obtain for my Observatory at Rivabellosa

	m	sec.
	11	43·7
	11	42·4
	11	41·9
the longitude . . . . West	11	42·7

Observations for Latitude.

July 12. The reduced zenith distance of the sun's upper limb, when on the meridian, gave as the latitude of Rivabellosa

N. 42° 42′ 37″

July 13. An observation of α Lyræ on the meridian and of *Polaris* out of the meridian, when reduced, gave for the latitude freed from error of the level,

N. 42° 42′ 19″.

July 14. An observation of the zenith distance of the sun's upper limb previous to the meridian passage at the hour angle 24° 5′ 15″ gave, when reduced, the latitude

N. 42° 41′ 48″.

July 20. The reduced zenith distance of the sun's upper limb, when on the meridian, gave the latitude

N. 42° 41′ 17″.

Combining the foregoing determinations of latitude, we obtain

	°	′	″
July 12. Observations of sun	42	42	37
July 13. Observations of stars	42	42	19
July 14. Observations of sun	42	41	48
July 20. Observations of sun	42	41	17
as the mean . . . . Latitude N.	42	42	

The foregoing numbers are not so accordant as I could desire; but they are, I believe, as good as could have been obtained with the instrument, particularly under the circumstance of its continual exposure to the sun during its employment in the day-time.

Elevation of the Station.

Before leaving Bilbao on the 10th, the aneroid barometer was read off, when it stood at 30·019 in., temperature 71° Fahr. On arriving at Rivabellosa it indicated 28·473 in., the temperature being 65° Fahr. With these numbers I estimated the height to be 1481 feet above the ground floor of Mr. VIGNOLES' house at Bilbao, which is several feet above the mean sea-level.

Mr. PRESTON, however, was so kind as to connect my station by levelling with a normal point on the railway, and made its height to be 1572 feet 4 inches above the mean sea-level.

Recapitulation.

The geographical position of my observatory at Rivabellosa was, therefore,

Latitude N. 42° 42′, Longitude W. 11 min. 42·7 sec.,

and its height above the mean high-water mark 1572 feet 4 inches. Mr. STRUVE has communicated to me that the geodetic connexion of Rivabellosa and Pobes showed the geographical position of my observatory to be

Latitude N. 42° 43′ 24″, Longitude W. 11 min. 41·3 sec.

Lastly, I estimate the error of the mean-time chronometer, Frodsham No. 3094, July 18th, 0^h, to have been 4·6 seconds fast of Greenwich mean time, by assuming a progressive increase in its losing rate from July 16th to July 18th, and taking the mean between Dr. WINNECKE'S and my own estimate for its error on July 16th.

After the return of the expedition, Mr. CARRINGTON kindly made some extensive calculations to admit of a direct comparison of my observed results with the demands of theory. To Mr. FARLEY I am also much indebted for special computations of the moon's position in respect of the sun's, in a form the most convenient for comparison with measurements hereafter to be mentioned.

Abstract of the Results of Mr. CARRINGTON'S *Calculations for Rivabellosa.*

Assumed position of station:—

Geographical latitude	42° 42′
Longitude W. of Greenwich	$11^m\ 42^s{\cdot}7$.
Height above sea-level	1572 feet.

Whence the following elements:—

Geocentric latitude	=42° 30′·5.
Log. distance from earth's centre . . .	9·9993076.

The true positions of the sun and moon are those of LE VERRIER'S and HANSEN'S

Tables, as derived from the Special Circular issued by the Superintendent of the Nautical Almanac.

For the totality the apparent positions were calculated for 3 h. 0 m. and 3 h. 5 m. Greenwich mean time, with the following results:—

Zenith distance of sun's centre	40° 33′.
Angle at sun between pole and zenith	47° 59′.
Moon's semidiameter	16′ 33″·0.
Sun's semidiameter	15′ 44″·8.
Totality began at 3 h. 0 m. 37·4 s. at 101° 38′ on the sun.	
Totality ended at 3 h. 4 m. 1·6 s. at 312° 56′	,,
Direction of motion from 297° 17′ to 117° 17′	,,
Relative motion during totality	92″·8.
Nearest approach of centres	13″·0.
Duration 3 min. 24·2 secs.	

For the first contact the apparent positions were calculated for 1 h. 40 m., 1 h. 50 m., and 2 h. 0 m. Greenwich mean time, with the following results:—

Zenith distance of sun's centre	28° 49′.
Angle at sun between pole and zenith	35° 50′.
Moon's semidiameter	16′ 34″·6.
First contact at 1 h. 47 m. 56·0 s. at 296° 54′ on the sun.	
Rate of approach of centres per minute . . .	24″·87.

For the last contact the apparent positions were calculated for 4 h. 0 m., 4 h. 10 m., and 4 h. 20 m. Greenwich mean time, with the following results:—

Zenith distance of sun's centre	52° 45′.
Angle at sun between pole and zenith	51° 30′.
Moon's semidiameter	16′ 30″·7.
Last contact at 4 h. 10 m. 15·2 s. at 117° 20′ on the sun.	
Rate of retreat of centres per minute	29″·85.

The formulæ used in computing the moon's parallax and apparent semidiameter were

$$\vartheta = \text{Æ} \text{ of zen.} \qquad m = \frac{\varrho \cos\phi' \sin p}{\cos\delta} \qquad n = \frac{\varrho \sin\phi' \sin p}{\sin\gamma}$$

$$\alpha' - \alpha = -\frac{1}{\sin 1''}\left\{m \,.\, \sin\overline{\vartheta - \alpha} + \tfrac{1}{2}\, m^2 \sin 2 \,.\, \overline{\vartheta - \alpha} + \tfrac{1}{3}\, m^3 \sin 3 \,.\, \overline{\vartheta - \alpha}\right\}$$

$$\delta' - \delta = -\frac{1}{\sin 1''}\left\{n \,.\, \sin\overline{\gamma - \delta} + \tfrac{1}{2}\, n^2 \sin 2 \,.\, \overline{\gamma - \delta} + \tfrac{1}{3}\, n^3 \sin 3 \,.\, \overline{\gamma - \delta}\right\}$$

$$\tan\gamma = \tan\phi' \,.\, \frac{\cos\frac{1}{2} \,.\, \overline{\alpha' - \alpha}}{\cos\left(\vartheta - \frac{1}{2}\overline{\alpha' + \alpha}\right)}.$$

$$R' = R \,.\, \sin(\gamma - \delta') \,.\, \operatorname{cosec}(\gamma - \delta).$$

Mr. Farley's *Elements of the Eclipse for Rivabellosa.*

The following calculations are based on the same latitude and longitude as those of Mr. Carrington.

Greenwich mean time.	Apparent distance of ☉ and ☾ centres.	Angle of line joining centres, N. towards E.	Augt. S. D. or Radius of ☾.	Ratio of Lunar to Solar radius.
d. h m	′ ″	° ′	′ ″	
July 18. 1 45	33 32·8	296 54	16 34·5	1·0526
1 55	29 22·0	296 54	16 34·4	1·0525
2 5	25 8·9	296 52	16 34·3	1·0524
2 15	20 53·0	296 47	16 34·1	1·0523
2 25	16 34·4	296 38	16 33·9	1·0521
2 35	12 12·2	296 23	16 33·7	1·0518
2 45	7 47·2	295 48	16 33·4	1·0515
2 55	3 19·3	293 40	16 33·1	1·0511
3 5	1 13·6	126 59	16 32·8	1·0508
3 15	5 48·2	119 20	16 32·5	1·0505
3 25	10 26·7	118 22	16 32·2	1·0502
3 35	15 9·1	117 57	16 31·9	1·0498
3 45	19 55·2	117 43	16 31·6	1·0495
3 55	24 45·0	117 32	16 31·3	1·0492
4 5	29 39·0	117 22	16 30·9	1·0488
4 15	34 37·2	117 14	16 30·5	1·0484

	h	m	sec.	
Time of first contact	1	47	57	at 296° 54′ N. towards E.
middle . .	3	2	20	duration of totality 3 min. 20 sec.
last contact	4	10	15	at 117° 18′ N. towards E.

Nearest approach of centres 0′ 12″·7.

OBSERVATIONS OF THE ECLIPSE.

1. *Observations with the unassisted Eye, and with the Telescope.*

A splendid day on Sunday the 15th was succeeded by one of the grandest and most awful thunder-storms I have ever witnessed; and the 16th was cloudy, almost without intermission. The day previous to the eclipse had been completely overcast, with the exception of a short interval about noon; but even then the sun could only just be seen through a cloud somewhat thinner than those which obscured the rest of the heavens. The climate had therefore proved anything but propitious, and every interval of fine weather had to be diligently made use of for the adjustment of the instruments and the prosecution of observations. Fortunately an opportunity had presented itself on two days for practice in observing the sun with the Dallmeyer between 1 h. 30 min. and 4 P.M., and for special practice at about 3 o'clock. It was ascertained that during the progress of the eclipse the radius bars would have to be changed from one leg of the tripod-stand to the other, and arrangements were made to prevent the necessity for doing this during or near the period of totality.

To this instrument I had fitted an eyepiece of my own contrivance, which I described, verbally, at one of the meetings of the Astronomical Society, and which was in consequence adopted by Mr. Pritchard. No account having been published of this appendage, and experience having proved its value in eclipse observations, I think it desirable to describe it here. It will be remembered that Mr. Hodgson some time ago proposed that a piece of polished glass should be used as a diagonal reflector in observing the sun; and "Hodgson's solar eyepiece" has been generally adopted, and is a most convenient and efficient instrument. It occurred to me, that if the glass reflector were made in the form of a parallelogram, of such dimensions that a moiety of its surface would suffice for the field of the telescope, one-half of the upper reflecting surface might be silvered and the other left plain, and that the addition of a suitable contrivance would enable the observer to draw into position the unsilvered or the silvered surface, according as either partial or total reflexion might be required. The silver film is so extremely thin that it in no way affects the focus, yet it is susceptible of the highest possible polish. It was not a convenient plan to silver only half the mirror; so, when the whole had been silvered *, one-half of the silver was neatly removed by means of a cloth, wetted with cyanide of potassium, strained over the forefinger. The roughened back of the reflector was freed from silver, and the plate then washed thoroughly with distilled water and allowed to dry. A little pad of wash-leather, well charged with dry rouge by rubbing it on a second piece of leather on which some rouge-powder had been placed, very soon removed the peach-like bloom from the silver surface, and produced a perfect polish.

The construction of the eyepiece will be readily understood by reference to the accompanying wood-engravings, wherein the same letter refers always to the same part.

Fig. 1

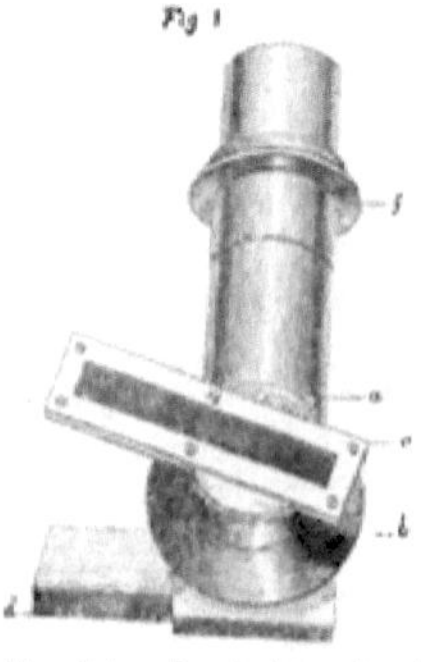

Fig. 2

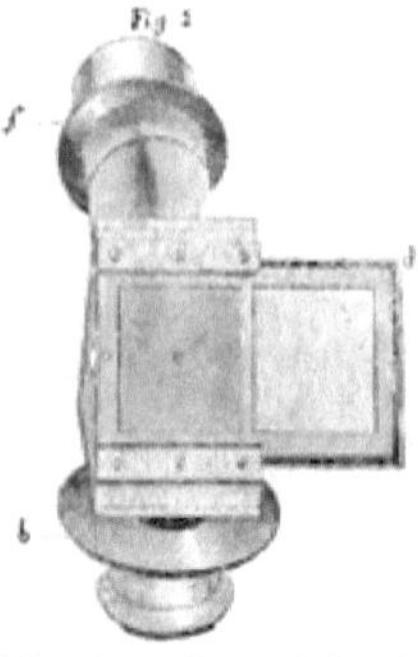

Fig. 3

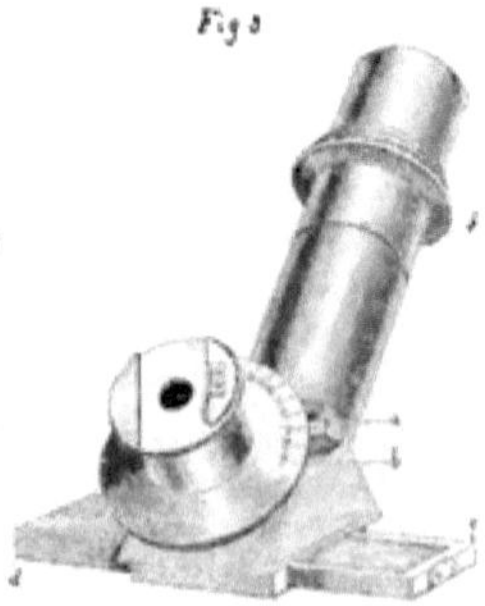

Fig. 1 is a front view, the plain glass reflector being in operation. Fig. 2 is a view of the under side, the plain glass being still in position for use. Fig. 3 shows the glass

* A method for silvering glass has been described by myself and Dr. Müller in the 'Monthly Notices of the Astronomical Society,' vol. xix. p. 171.

reflector drawn out so as to place the silvered surface in the field. *f* is the socket-adapter, which screws into the telescope; it has a slit cut into it to receive a pin fixed on the sliding tube, the object of which is to keep the position-lines of the eyepiece, when once adjusted, in their proper position; *e* is the glass reflector, the half towards *d* being silvered, that towards *e* plain; *d* is a covering to protect the sliding reflector; *b* is a circle fixed to the body of the eyepiece, having one quadrant divided into nine spaces of 10° each; *a* is an index attached to a positive eyepiece, which can be moved by it through an arc of 90°; *c* is a graduated sun-shade, composed of a wedge of dark glass and another of white glass, reversed in position, so as to form, when combined, a parallel plate. This is held firmly in its place by means of a spring, shown in fig. 3, which, while it holds the shade firmly in any required position, also allows its instant removal at pleasure. The glass reflector *e*, as soon as the observer desires to use the silvered surface, can be drawn forward in a small fraction of a second, without disturbing any other part of the instrument.

In the focus of the positive eyepiece was fixed a piece of parallel glass on which were etched several lines; this micrometer-plate was carried round with the eyepiece whenever the index, *a* (figs. 1 & 3), was moved. A reference to Plate VI., which contains a fac-simile of my hand-drawings, and also a representation of the position-lines, will render clear the following explanation. Four principal lines on the glass plate formed a tangential square calculated to enclose exactly the moon's disk, which in fact it accomplished with great precision; four other lines surrounded the first square at the distance of exactly 1′ of arc; and a third series formed a third square at the same distance from the second. Joining the angles of the squares were two diagonal fainter lines, which served to measure angles of position, while the several squares served to measure distances. The angles of the tangential square may be designated A, B, C, D. As

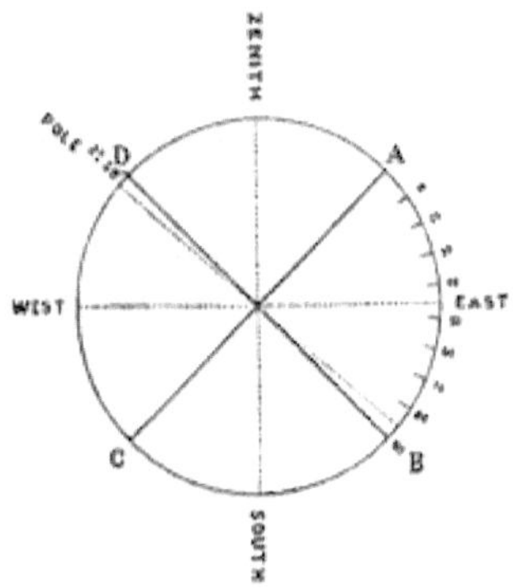

soon as the axis of the telescope-stand had been adjusted in a vertical position by means of screws affixed to the feet of the tripod-stand, and of a level attached to the vertical axis of the telescope, a distant mountain peak was made to run along one of the lines D A or C B by causing the telescope to turn on the vertical axis when the index of the eyepiece stood at zero; and the position of the whole eyepiece was regulated by means of a slight axial adjustment of the sliding tube until the line in question, and the course of the object horizontally across the field, showed an absolute coincidence. The diagonal lines D and A were then each 45° from the zenith; and the angle between the pole and the zenith at the period of totality being 47° 59′, the wire D, to the west of the zenith, was 2° 59′ east of the pole. This will readily be seen by the annexed diagram, which shows the apparent positions of the zenith and of the east, west, and south points in the field of the eyepiece, the zenith being represented in its natural position, but the east and west points reversed, right for left; so that, in referring any measurement to D, 2° 59′ must be added to reduce them to positions from the north point towards the east.

At last the eventful 18th of July arrived, and appeared hopelessly cloudy. The sky was watched with the most intense anxiety by us all; and I am free to confess that *my* nerves were in the most feverish state of agitation. Not the slightest break in the clouds or mist was visible until about 10 o'clock, when a streak of clear sky gave us the first faint gleam of hope. At noon the sky began to clear very generally, not so much by the clouds being carried off by the wind, as by their melting away in the air. An attempt was made to get an observation of the sun; but the clouds were so dense that it was only just as he was passing away from the field of the telescope that his following limb could be made out. Soon after this the clouds, which had been dissolving and gradually becoming thinner, disappeared all at once; and we had a magnificent sky, absolutely cloudless, except near the horizon. The heights towards the north in the direction of Pobes were perfectly free from mist; moreover we could discern through the telescope Mr. Bonomi and Mr. J. Beck on a low hill a few miles to the west of our station, and it gave us pleasure to know that at least two of the several parties were as fortunate as ourselves. The mists surrounded the higher peaks of the Pyrenees so pertinaciously that we expressed to each other great fears for our good friend Mr. Vignoles; for we were aware that he intended to carry out his plan of observations from one of the highest peaks, and we afterwards heard with the greatest regret that our apprehensions had been realized.

About twenty minutes before the commencement of the eclipse, an occurrence took place which very nearly brought all our labours to a calamitous termination. Mr. Preston had placed at our disposal his excellent and handy servant Juan, whom we had always found obliging, and very ingenious in expedients whenever any temporary arrangements had to be made. In order that he might have an opportunity of looking at the eclipse, I smoked a piece of glass for him with a wax lucifer-match, and he then, on his own account, prepared several pieces for the bystanders in a similar way. The demand soon increased so much, that he was scarcely able to keep pace with it, and at length became so excited that he threw away the matches in all directions without extinguishing them, and some, falling in the standing corn, set it on fire. The corn was very thin, and fortunately the wind was blowing from the position of the fire towards the thrashing-floor; otherwise but a few minutes only could have elapsed before the conflagration would have assumed such dimensions as to be beyond the power of man to control. Happily, a few seconds after the occurrence, the crackling sound and the smell of burning straw drew my attention to the spot, and, water being at hand, the fire was got under before it had spread more than a few feet.

The Alcalde of Miranda had intimated to me, a few days before, that he was instructed to place at my disposal as many of the civic guard as I might think necessary to prevent interruption; but my experience of the consideration evinced by the Spaniards was such, that I replied that one or two would be quite sufficient. Shortly before the commencement of the eclipse, there arrived five mounted guards, who were of great use in preventing the crowd from encroaching on the thrashing-floor, which an excusable curiosity to watch our proceedings tempted them often to do. It is right to add that I could not persuade the guards to take any present whatever, their reply being that their orders on this head were imperative, and that, moreover, they had felt pleasure in being of service. When we were on the point of commencing our observations, about 200 persons had assembled round our observatory, and, although they conducted themselves perfectly well in other respects, their talking quite prevented my hearing the beats of the chronometer. They seemed to think that the eclipse could only be seen from my station; and it was with some difficulty that a number were persuaded to go to an adjoining height, whence the effects on the landscape and the progress of the shadow could really be better observed. I explained this, through the kindness of a gentleman from Miranda who spoke French, and who showed his faith in what I stated by leading the way. The Alcalde of Rivabellosa, Cirilo Guinea, to whom I was indebted for facilitating my operations, explained to those who remained around the station the necessity for silence, and they thenceforth carried on their conversation in a tone which caused us no inconvenience. It is indeed impossible to speak too highly of the good feeling manifested throughout by the Spaniards of all grades, who endeavoured in every way to promote our objects.

Owing to my pocket chronometer having tripped, and become many minutes fast of Greenwich mean time, some confusion arose about the period of first contact, and a photograph, which it was intended to procure as close as possible to that event, was lost in consequence of the plates being prepared too soon, and none being ready when it actually took place. The error of the pocket chronometer was only discovered when it was too late, and it was then found to be faster than Frodsham 3094 by 8 min. 11 sec.

The first contact was observed at 1 h. 56 min. 55 beats.

	m	sec.	h	m	sec.
5 beats to 2 seconds . .			1	56	22
Error, fast of Frodsham	8	11			
Error, Frodsham fast of Greenwich mean time . . .		4·4		8	15·4
Making the observed Greenwich mean time of first contact to be			1	48	6·6

	h	m	sec.
The occultation of the largest solar spot, which I call *c*, occurred at 2 h. 10 m. 125 beats	=2	10	50
Deducting the difference between the pocket Frodsham and Frodsham 3094		8	11
It gives	2	2	39

as the time by Frodsham 3094. This agrees within 0·5 second of the time noted by Mr. Beckley when a photograph of the phenomenon was taken in accordance with my signal, namely, 2 h. 2 m. 39·5 secs. Attention is called to the very exact accordance of the times recorded by myself and by Mr. Beckley, because I shall hereafter have to draw attention to certain conclusions which I have deduced, the soundness of which is dependent in part upon the epochs of the photographs having been accurately registered by Mr. Beckley.

After the occultation of the spot *c*, nothing worthy of record occurred until about 2 h. 12 m., when a cluster of clouds formed very rapidly and unexpectedly in the immediate neighbourhood of the sun, and completely put a stop to both optical and photographic observations. The clouds melted away about six minutes after they had formed, and thenceforward until the end of the eclipse all went on without interruption.

I had never before witnessed so great an obscuration of the sun as that presented by this eclipse many minutes even before the totality occurred, and I was particularly struck by the change of colour in the sky, which had been gradually losing its azure blue and assuming an indigo tint, while at the same time I remarked that the surrounding landscape was becoming tinged with a bronze hue, which to my mind suggested the idea that the light of the sun near the periphery is not only less intense than, but possibly different in quality from, that of the centre*. Spectrum experiments at future great eclipses, when the sun's crescent is reduced to an extremely narrow line, would set this question at rest, and might also have an important bearing on the line of investigations so ably inaugurated by Kirchhoff and Bunsen, into the composition of the sun's atmosphere. Another phenomenon could not fail to attract attention. When the sun's visible disk was reduced to a very narrow crescent, the shadows of all near objects became extremely black and sharply defined, whilst the lights, by contrast, assumed a peculiarly vivid intensity, the aspect of nature strongly recalling to mind the effects produced by the illumination of the electric light. Several minutes before the totality, the whole contour of the brown-looking lunar disk could be distinctly seen in the heavens.

Only a few brief seconds, unfortunately, could be spared from the telescope after the totality had actually commenced; but when I had once turned my eyes on the moon encircled by the glorious corona, then on the novel and grand spectacle presented by the surrounding landscape, and had taken a hurried look at the wonderful appearance

* In connexion with this remark, compare Sir John Herschel on the Chemical Rays of the Spectrum. Philosophical Transactions, 1840, Art. 82.

of the heavens, so unlike anything I had ever before witnessed, I was so completely enthralled, that I had to exercise the utmost self-control to tear myself away from a scene at once so impressive and magnificent, and it was with a feeling of regret that I turned aside to resume my self-imposed duties. I well remember that I wished I had not encumbered myself with apparatus, and I mentally registered a vow, that, if a future opportunity ever presented itself for my observing a total eclipse, I would give up all idea of making astronomical observations, and devote myself to that full enjoyment of the spectacle which can only be obtained by the mere gazer.

Although, possibly, not more than twenty seconds were devoted to observations with the unassisted eye, the phenomena remain strongly impressed on my memory, and at the time of writing this account, sixteen months after the event, I have it now pictured before me mentally, as vividly as if it had but just occurred. The darkness was not nearly so great as I had been led to expect from the accounts which I had read of former total eclipses; and although I had a lantern at hand, I did not require it, either in making my drawings or for reading the divisions of the micrometer quadrant on the eyepiece. The illumination was markedly distinct from that which occurs in nature on any other occasion, and certainly was greater than on the brightest moonlight night; and yet, at the time, the light appeared to me less than what I remembered of bright moonlight. It was only by subsequent trials, in endeavouring to make out details of the drawings which I had made of the phenomena, and to distinguish between colours under various circumstances of moonlight and twilight, that I was able to form a proper appreciation of the amount of light; and the best account I can give of it is, that it most resembles that degree of illumination which exists in a clear sky soon after sunset, when after having made out a first-magnitude star, other stars of less brilliancy can be discerned one after another. The light was good enough and sufficiently polychromatic to enable me to distinguish the colours of near objects; but those in the distance appeared to be illumined by the most unearthly hues.

Immediately surrounding the corona, the sky had an indigo tint, which extended to within about thirty or twenty-five degrees of the horizon, while lower down it appeared to me to be modified by a tinge of sepia. It became red as it approached the horizon, close to which, and just above the mountains, it was of a brilliant orange. The mountains appeared, by contrast, of an intensely dark yet brilliant blue. I saw two stars to the east of the sun, which by the aid of Mr. Hind's diagram I have since identified as Jupiter and Venus; but I had not time to search for more, or, most probably, I should have seen others. These planets, and also Castor, were made out by Messrs. Robert Swanson, Harry Edmonds, and Matthews, in the employ of Messrs. Brassey and Co., and were identified by them in my copy of Mr. Hind's diagram.

The effect of totality upon the bystanders was most remarkable. Until the beginning of totality, the murmur of the conversation of many tongues had filled the air; but then in a moment every voice was hushed, and the stillness was so sudden as to be perfectly startling; then the ear caught the sound of the village bells, which had been tolling unheeded during the eclipse, and this circumstance added much to the solemn grandeur of the occasion.

The time I could spare was far too short for any exact observations of the corona; however, I knew that Mr. PRITCHARD, Mr. OOM, Mr. BONOMI, and other observers intended to make special delineations and measurements of that phenomenon, and I therefore confined my attention to its general characteristics. It appeared to me to glow with a silvery-white light, softening off into a very irregular outline, while from its general boundary shot out several long streamers. It extended generally to about 0·7 or 0·8 of the moon's diameter beyond her periphery. Close to the moon, and reaching not further than 2′, the light was very brilliant, and several zones of gradations of brightness appeared to exist, but the very bright zones would all be comprised in a circle about 0·25 of the moon's diameter.

The observations just recited were made in the brief interval I could afford between the telescopic observations, which I will now proceed to describe. In order to facilitate my operations, I had prepared two diagrams exactly representing the appearance of the micrometer lines in the telescope, and, by chance, I had made the tangential square of such dimensions as to include a circle precisely 4 inches in diameter, which had been coloured to render it more readily distinguishable, and which represented the moon. Four inches happened to be almost exactly the diameter of the moon's disk on the screen of the heliograph; so that, later on, the photographs and the two drawings made during the totality, were readily compared by the superposition of each upon each.

On the diagrams I had painted fifteen streaks of various tints, some of which I believed might resemble the colour of the prominences, and some I knew would be useful as a contrast, to enable the eye to form a more correct judgment. The chromatic scale I here insert contains a selection of the tints painted round my diagram.

SCALE OF COLOURS WITH WHICH THE PROMINENCES WERE COMPARED.

Several minutes (probably five) before totality, I entirely removed the dark glass, and found that the sun's image might be looked at without the slightest inconvenience after reflexion by the plain glass. I could then see in the telescope, as I had shortly before seen with the naked eye, the whole of the lunar disk, which appeared of a deep sepia brown, nearly, but not quite, black, and, to my great surprise, I perceived a luminous prominence, about 20° to the west of the zenith, shining with great brilliancy, although, on account of the plain-glass surface being then in use, the greater part of its light passed through the glass, and was therefore not reflected to the eye. I then, cautiously, but rather quickly, brought into action the silvered surface, and beheld with delight that the luminosity of the prominence, which I will call A*, was so great that there could be very little doubt of our obtaining the much wished-for photographic pictures.

I now watched carefully for the so-called Baily's Beads, but no such phenomena presented themselves,—at which, however, I felt no surprise, for I had always believed that they arose, in all probability, from the atmospheric disturbance of an image formed by a telescope wanting in perfect definition. The Dallmeyer I used was so perfect that I did not think I should see anything of the kind.

To the east of the zenith, about 20°, a floating cloud, quite detached from the moon's limb, and distant from it more than 0'·5, next attracted my attention. This cloud, which I will call C, appeared about 1'·5 long, and was inclined about 50° or 60° to the moon's limb. It had two curvatures, both convex on the edge most distant from the moon, and was decidedly of a rose tint, but of a much paler hue than the published accounts of previous eclipses had led me to expect. I compared the prominence carefully with my scale of tints, and found that it very nearly matched the colour marked *c*. It must therefore have been of a yellowish pink (approaching a salmon); for *c* on my chromatic scale was a mixture of carmine and cadmium yellow. This prominence (C) presented a great amount of detail, and reminded me of the aspect of a cirrus cloud glowing with the illumination of a setting sun. I should here remark that, in comparing my scale of colours with the luminous prominences, I depended on the general light of the heavens, and that I did not employ my lamp, which, I found, completely changed their appearance.

The prominence A was generally more brilliant, and did not seem to me to be so pink as the detached cloud; I could, moreover, detect a tinge of yellow in its brilliant light. It also showed considerable structure, appearing to consist of several streaks, curved inwards, while from the summit came two peach-coloured faint streamers, bending over in opposite directions downwards towards the moon's limb.

I paid most particular attention to the prominence A, because I knew from its position that it was critically placed for the observation of any change of position-angle in reference to the moon's centre; and I also remarked carefully the prominence C, and sketched all that I could make out by the most careful scrutiny. On comparing my drawings with the photographs, it will be perceived that a certain boomerang-like prominence in the photograph is wanting in my hand-drawings, and that there are also other prominences visible in the photograph which are not shown in the drawings. This is a curious circumstance, hereafter to be more particularly dwelt upon; but it is right to mention it here, because it affords me the opportunity of saying that, at all events, as regards the boomerang, I am certain that it was not visible in the telescope; for I observed so carefully in the neighbourhood of the floating cloud, that it is next to an impossibility that such an object could have escaped detection.

* See Index Map, Plate XV.

In the eastern quadrant (in reference to the zenith) a long line of prominences, extending over 70° on the moon's limb, was visible at the commencement of the totality, but before the end of totality it was covered by the lunar disk. This streak (which I call I) terminated in a hook, about 1′·5 high, bent upwards, and was *much indented on the concave side*, where it was in contact with the moon's edge. It was extremely brilliant, and, although it presented in parts a pink colour, was not uniformly so coloured, but to my eye had here and there a considerable admixture of yellowish white. In the first photograph of the totality is depicted a curious branching prominence, not unlike the fallen stump of a tree, which I did not observe, and therefore did not record in the sketches. I do not state so positively that this prominence was not visible, for this reason, that I did not pay such special attention to that part of the field, my eye being directed more particularly to the prominences A and C; but I have a strong impression that it was not visible. Just about the part where this would be, the corona appeared to me, in the telescope, to be particularly bright; but, besides a mere sheet of brilliant light, I saw nothing to delineate.

About half a minute after the commencement of totality, the progress of the moon uncovered, in the western quadrant (in reference to the zenith), a small peak, like that of a mountain, which I will call R. As the eclipse progressed, this prominence became more and more uncovered, and another smaller peak appeared, the whole contour reminding me somewhat of the hull and masts of a ship in full sail. Just before the reappearance of the sun this prominence reached apparently about 1′·5 beyond the moon's limb.

Extending from the southern base of this prominence, there came into view, about a minute before the end of totality, a long streak of prominences much *indented and irregular on the concave side*. This streak extended over fully 50° on the moon's limb, when it had been fully uncovered by her onward course. It was pretty generally of a decided rose tint. Just previously to the reappearance of the sun, I remarked a sort of carmine glow near that part of the moon's limb where the crescent of the sun was first re-formed.

Plate VI. is a most exact fac-simile of the two drawings, black representing white, which I made during the totality, and it is desirable that I should make a few remarks about them.

Figure 1 was begun, as nearly as I can recollect, about thirty seconds after the commencement of totality. As a preliminary step, the moon's disk was brought exactly within the tangential square, and the position of the prominence A, in respect of the line D, was noted first of all, and at once marked down on the left-hand diagram; the hook I was then referred to the line B, and the mountain R to the line D, the latter being registered with great care. The floating cloud and the other prominences were then filled in, possibly not quite so carefully. The details were next drawn in, black representing white, and the first diagram was completed as rapidly as possible, yet as faithfully as the short time at command would permit. I was aware, whilst so occupied, that by the addition of detail after detail in the several prominences I was exaggerating their dimensions; but there was too little time to spare to rub out and commence anew.

When the first drawing was completed, about a minute and a half after the commencement of the totality, I looked away from the telescope in order to make the eye observations which I have already described, and before I resumed my work at the telescope an interval of half a minute may have elapsed, but certainly not more. The next thing I did was to measure the angular position of the prominence A; and after bringing the moon well into the tangential square, I moved the wire D through the arc necessary to bring it into contact with the side of that prominence nearest to D, which brought the index to an exact coincidence with one of the divisions on the quadrant; I noted down 10° for the angle moved through; but this is an evident error, for the angle was as nearly as possible 20°, which, added to 2° 59′, makes the position-angle of the western boundary of that prominence 22° 59′, from the north towards east, which is not far from its true position at that time.

Whilst measuring this prominence, I asked Mr. Reynolds, whose allotted task it was to develope the photographs after their exposure in the heliograph, whether anything could be seen on the first plate of the totality; and learning, with a thrill of intense pleasure, that the operation had completely succeeded, I made no further measurements, knowing full well that I should get them far better in the photographs.

Immediately after this, I commenced my second drawing, given in Plate VI., and noted down the position of the prominences A and R very exactly, by referring them to position-line D; and I then filled in the other details. As very little time remained for the completion of the drawing, I devoted my attention chiefly to the prominence R and a faint hooked prominence about 45° to the west of the position-line D, which did not imprint its image on the second photograph to the extent I should have expected from its dimensions in my sketch.

Between the completion of the first sketch and the commencement of the second, I estimate that there was an interval of about one minute, and that the second sketch was therefore commenced as nearly as possible 2½ minutes after the beginning of totality.

Thus, before commencing sketch No. 1, there elapsed,

	min.	sec.
From the beginning of total obscuration	0	30
To complete No. 1 sketch it required	1	0
Time consumed by eye observations, away from the telescope	0	30
The measurement of prominence A occupied	0	30
Interval elapsed from the beginning of totality to the commencement of sketch No. 2	2	30

By placing a horn protractor on the original sketches, the following measurements were made:—

Protuberance.	Synonym.	Part measured.	Distance from position line D towards east, in degrees and decimals of a degree.	
			First drawing.	Second drawing.
A.	Cauliflower. Wheatsheaf.	First boundary, *a* *	23·0	20·0
		Second boundary, *a'*	28·0	24·5
		Middle	25·5	22·25
C.	Detached cloud.	First point, *c*	58·7	49·0
		Last point, *c'*	69·0	58·5
R.	Mountain peak.	First point, *r'*	347·0	343·5

In order to reduce these measures to position angles from the North towards East, it is necessary to add to them 2° 59', say 3°, which will give us—

	First drawing.	Second drawing.	Apparent angular motion in the interval.
A. Middle	28·5	25·25	3·25
C. First point, *c*	61·7	52·0	9·7
R. First point, *r'*	350·0	346·5	3·5

As I before stated, the positions of A and R were laid down with great care; and it will be hereafter seen that their deviations from the positions given by measurements of the photograph are remarkably small. The measurement of all the details, however, do not agree so well, because the same care could not be devoted to the laying down of their positions.

These drawings show that there was a decided angular shifting of the luminous prominence A, and of others, in reference to the moon's centre; and taking into account the probable interval between the two drawings, namely two minutes, the amount of angular motion of A is a very near approximation to the angular change which must actually have occurred. As mentioned above, there is in the drawings an exaggeration of the dimensions of the prominences, which renders them unfit for the precise determination of the moon's actual progression in the line of motion during the period of totality; nevertheless they afford excellent evidence that there was, in fact, a covering and an uncovering of prominences, which, taken in connexion with the change in the position-angle of the protuberance A with reference to the moon's centre, can only be explained on the assumption that these extraordinary appendages belong to the sun, and not to the moon.

Furthermore, it would be quite possible to make out, with considerable although not with absolute accuracy, from these drawings, the direction of the moon's motion, and the extent to which the prominences first seen were obscured by the progress of the lunar disk, and others uncovered on the opposite side as the moon continued her course. For instance, it will be remarked on inspection, that the streak of prominences, almost 1' in

* The letters refer to the index map, Plate XV.

height, depicted in the eastern quadrant of fig. 1, Plate VI., is almost entirely covered in fig. 2, and that the difference of position of the moon in the two pictures, when measured by a suitable scale, indicates a motion of about 50" in the interval of two minutes which they include,—a result very near to the truth, for the actual progression in that period was 54"·5. The photographs, however, as will be hereafter seen, are so much better adapted for such determinations as these, that it is not worth while to dwell more upon the conclusions to be derived from the hand drawings.

In the two coloured drawings, Plates VII. and VIII., I have depicted the result of my telescopic observations; to facilitate my doing which at some future convenient time, I made a coloured sketch on the afternoon of the eclipse. This coloured sketch, the black-and-white drawings made at the telescope and shown in fac-simile in Plate VI., together with my photographs, which I have not hesitated to use to correct any errors of position or dimension in the sketches, have enabled me to give in these drawings what I believe to be a very truthful representation of the appearance of the prominences, immediately after the commencement and just before the end of totality. The corona I do not give as an absolutely true representation of that phenomenon, but as fairly resembling its general appearance. It has been derived from the photographs, so far as they show it.

II. *Account of the Photographic Observations.*

The Kew heliograph, with which the photographs were obtained, is represented in the accompanying engraving *. It was devised by myself, for the special object of making photographs of the sun's disk, at the request of the Council of the Royal Society, in accordance with a recommendation to that effect by Sir JOHN HERSCHEL.

It has an equatorial mounting of the ordinary form, after the so-called German model, to which is attached a clock-work driver. The tube is square in section, and larger at the lower end than at the upper or object-glass end. The object-glass has 3·4 inches' clear aperture and 50 inches' focal length; the primary focal image of the sun at his mean distance is 0·466 inch in diameter; but before it is allowed to fall on the sensitive plate, it is enlarged to about 3·8 inches by means of an ordinary Huyghenian eyepiece. In the plane of the focus of the posterior lens of this eyepiece are attached two position-wires, which cross at right angles, and which were adjusted, approximately, into a position at an angle of 45° to a parallel of declination. The object-glass is so constructed as to ensure the coincidence of the chemical and visual foci; but this coincidence being somewhat disturbed by the Huyghenian secondary magnifier, the amount of adjustment required to effectuate the best chemical focus was ascertained very carefully by a series of experiments.

* The engraving was copied from a photograph taken at Rivabellosa. The front boards of the observatory were taken out in order that this might be done. Mr. DOWNES, who was charged with the preparation of the plates, is standing in the doorway leading to the developing room. Mr. REYNOLDS has a plate-holder ready to place in the heliograph, and Mr. BECKLEY is observing the time with the chronometer.

For sun-pictures, and the photographs of the several phases of the eclipse, the aperture of the object-glass was reduced to about 2 inches in diameter by means of a stop; but the light of the sun is so extremely powerful, that, even with this small aperture, combined with the enlargement of the primary image and its consequent reduction in intensity by 64 times, the shortest exposure possible with the ordinary means of uncovering and covering the object-glass would be far too long, and would give none but solarized pictures. For this reason the instrument has attached to it an instanta-

neous apparatus of a peculiar construction. It consists of a sliding plate with a square aperture sufficiently large to permit of the passage of all the rays; this aperture is fitted with a sliding piece, actuated by a screw which projects through and a few inches beyond the telescope-tube; by means of this screw the aperture may be completely opened, closed, or reduced to a slit of any required width; a divided scale being affixed to the screw for that purpose. The projecting screw connected with the slide is shown in the engraving, on the underside of the tube.

Previous to taking the picture, the sliding plate is drawn up just so high that the unperforated part of it completely shuts off the sun's image; it is held in this position by means of a small thread attached to it at one end and looped at the other, the loop being passed over a hook on the top of the tube; and the slide is pulled downwards, in opposition to the thread, by means of a spring of vulcanized caoutchouc attached to the inferior side of the tube. When the picture is about to be taken, the retaining thread is set on fire*, and the rectangular aperture, as soon as the sliding plate becomes released, flashes across the axis of the secondary object-glass—thus allowing the different parts of the sun's image to pass through it in succession, and to depict themselves one after another, after enlargement, on the collodion-plate. Although the time of exposure is so short as to be scarcely appreciable, yet it is necessary to regulate its duration; and it is therefore controlled by adjusting, 1st, the strength of the vulcanized caoutchouc spring; 2ndly, the width of the aperture. In practice, the opening is usually varied between one-twentieth and one-fortieth of an inch.

A number of plates, with ground rims and edges, were cleaned in London, so as to permit of their examination, and all defective ones were rejected; forty-eight selected plates were then numbered consecutively, and arranged in boxes marked very distinctly A, B, C, D, so as to ensure their being taken out in the proper order during the eclipse. The heliograph was furnished with three plate-holders, in order that no interruption might occur in the succession of the photographs; and as these were filled, they were placed in such a way that each plate was sure to be exposed in its numerical order. A few spare plates were also cleaned, and marked A, B, C, D, E, F, G, H, I, &c.

On the day previous to the eclipse the plates were again carefully cleaned, and replaced in their proper order in their respective boxes.

On the 18th the following plates were placed in the heliograph, and the time of taking each photograph noted by Mr. BECKLEY, with any requisite remarks. The time was observed with Frodsham No. 3094, whose error at Greenwich mean noon was, as already stated, fast of Greenwich mean time 4·6 seconds, and whose daily rate was losing 2·1 seconds. The exact time of depiction was ascertained by listening to the click which the instantaneous slide made in striking home upon a stop, when it had flashed across, in front of the secondary magnifier.

* Mr. CLARK, who undertook this task, is represented in the engraving with a lighted taper in his hand.

No. or letter.	h	m	sec./2	Remarks.
1.	23	38	1	Through a cloud.
2.	....			Spoiled.
3.	....			Spoiled.
4.	0	23	105	
A.	0	29	45	
B.	0	34	34	
C.	0	42	24	
D.	0	49	93	
E.	1	3	23	
F.	1	13	85	
G.	1	25	45	
H.	1	29	76	
I.	1	41	76	
5.	1	45	3	
6.	1	47	96	
7.	1	52	71	The first of the eclipse.
8.	1	56	15	
9.	2	2	79	Occultation of spot *c*.
10.	2	3	41	
11.	2	7	21	
12.	2	11	31	
13.	....			Clouds.
14.	2	20	24	
15.	2	22	97	
16.	2	27	103	
17.	2	33	36	
18.	....			Spoiled.
19.	2	36	43	
20.	2	41	55	
21.	2	46	46	
22.	2	48	60	Occultation of spot *b*.
23.	2	51	53	A solarized picture, in consequence of the full aperture being used and the instantaneous slide being detached.
24.	2	55	88	Id., the time uncertain.
25.	....			Totality; time not noted.
26.	....			Totality; time not noted.
27.	....			Shaken by wind; time not noted; the instantaneous slide had not yet been replaced.
28.	3	13	116	
29.	3	17	32	Reappearance of spot *c*.
30.	3	20	24	
31.	3	24	35	
32.	3	26	112	
33.	3	34	12	
34.	....			Spoiled.
35.	3	37	48	
36.	3	41	46	
37.	3	44	4	
38.	....			Forgot to uncover the plate.
39.	3	51	116	Spot *a* uncovered completely.
40.	....			Forgot to uncover the plate.
41.	3	59	5	Reappearance of spot *b*.
42.	4	2	96	
43.	4	5	98	
44.	4	10	75	
45.	4	16	40	

The diagram shows the appearance of the cross wires when projected on the glass screen. The image of the sun, being twice reversed, is finally depicted on the screen in its natural position, north being at the top, south at the bottom, east to the left hand, and west to the right hand. In the positive photographs of the eclipse, printed from the negatives, the pictures are likewise erect, and the points similarly situated. Calling the wires I., II., III., IV., I. would have approximately the position-angle of 45°, II. 135°, III. 225°, and IV. 315°. In the measurements, hereafter to be described, the several photographs were so placed on the measuring instrument as to cause its circle to read respectively one or other of these angles, according as either I., II., III., or IV. was employed in adjusting the photograph, the correction to the measured angles, necessitated by the deviation of that wire from its assumed position in reference to a parallel of declination, being subsequently applied.

The wires were found not to be absolutely at right angles.

	°	′
IV.— I. measured	89	59·3
I.— II.	89	52
II.—III.	90	8·7
III.—IV.	90	
	360	0

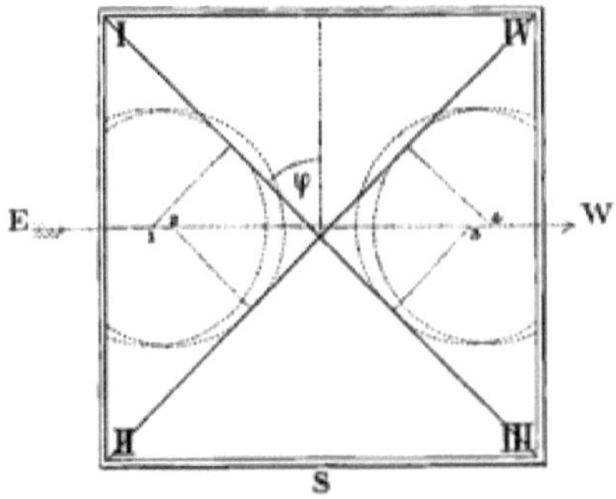

At 4h 35m, when the heliograph was pointed to the west of the meridian, observations were made to determine the deviation of the position-wires, from an angle of 45° to a parallel of declination, by the method described by Mr. Carrington in the 'Monthly Notices' of the Astronomical Society, vol. xiv. p. 153; and the observations were repeated on July 19 at 11h 55m, when the heliograph was pointed east of the meridian.

July 18, 4h 35m, by Frodsham 3094 uncorrected.

The sun's limb made contacts with wires I. and III.

	sec.
at an interval of	194
" "	195
" "	193
Sum	582

and with wires II. and IV.

at an interval of	186·5
" "	191
" "	188
	567·5

582 log = 3·7649

567·5 log = 3·7540

Angle of φ 45° 43′·5 log tan. = 0·0109

July 19, 11h 55m.

The sun's limb made contacts with wires I. and III.

	sec.
at an interval of	189·5
" "	188
Sum	377·5

and with wires II. and IV.

at an interval of	. . . .	193·5
" "	. . . .	193
Sum	. . . .	386·5

whence angle $\phi = 44° \; 19'·5$.

	h min.			° ′
At	4 35	P.M.	$\phi =$	45 43·5
At	11 55	A.M.	$\phi =$	44 19·5
Interval	4 40	diff.	+	1 24·0
Interval . . . . . =	280		=+	84

$\frac{+84}{280} = +0'·3$ the change of the angle ϕ in 1 minute.

	h	m	sec.
No. 6 photograph was taken at	1	47	48
Deducting	11	55	0
	1	52	48 = 112·8 min.

$112·8 \times 0'·3 = 34'$, $44° \; 19'·5 + 34' = 4° \; 53'·5$, the position of wire I. in reference to a circle of R.A. at the epoch of No. 6 photograph; hence the correction to be applied to the assumed position of 45° is $-0'·5$.

With the foregoing data has been calculated the following Table of corrections, to be applied to the assumed position of the wires at the epochs of the several photographs.

No.	Correction to the wires.	No.	Correction to the wires.	No.	Correction to the wires.
6.	−6′·5	20.	+ 9′·5	32.	+23′·2
7.	−5·1	21.	+11·0	33.	+25·4
8.	−4·0	22.	+11·7	35.	+26·4
9.	−2·0	23.	+12·6	36.	+27·6
10.	−1·8	24.	+14·0	37.	+28·4
11.	−0·7	25.	+15·5	39.	+29·8
12.	+0·5	26.	+16·5	41.	+31·9
14.	+3·2	27.	+18	42.	+33·0
15.	+4·0	28.	+19·4	43.	+33·9
16.	+5·5	29.	+20·3	44.	+35·3
17.	+7·1	30.	+21·2	45.	+37·0
19.	+8·0	31.	+22·5		

In these corrections I have not taken into account a small error in the computed angle of ϕ, which arises in consequence of the wires not being at right angles; for, on examining them, I found that the heat of the sun had caused a curvature in one, and I could not, without much trouble, have ascertained the correction with precision. It was computed that it would not, however, amount in any case to more than $+2'$. My measurements of position-angle, hereinafter given, are moreover liable, from the difficulty of adjusting the photographs, to discordances to the like extent, as will be easily conceived when it is stated that 2′ on the sun's limb do not occupy more than the space of $\frac{1}{1000}$th of an inch on the photographs obtained with the Kew heliograph.

At the moment of taking the photographs, the collodion was in a soft and moist condition, but subsequently, when the measurements were made, it had become dry. It became, therefore, not only a matter of interest, but of fundamental importance, to ascertain whether there had been any contraction of the collodion while drying, and, if so, whether the contraction had been uniform. Much care and attention were necessary in order to determine this point. By observing the positions of specks on the glass in respect of markings on a photograph while wet, it could be seen whether they retained their relative positions when the collodion had dried. The result, however, proved that there was no appreciable contraction, except in thickness, and that the collodion film did not become distorted, provided the rims of the glass plate had been well ground. I cannot show this more strikingly than by citing the measured radius of the sun on two photographs, namely, Nos. 6 and 45, and the measurement of the angles between the position-wires depicted on them. The radius of the sun No. 6 was found to be 1906·5, that of No. 45 1906·0 thousandths of an inch.

	Angle between IV. and I.	Angle between I. and II.	Angle between II. and III.	Angle between III. and IV.
No. 6	89° 59′·3	89° 52′	90° 8′·7	90° 0′
No. 45	90 2	89 53	90 6	89 58·5
Diff. 6—45 . . .	−2·7	− 1	+ 2·7	+ 1·5

These differences, which are extremely small, do not exceed those obtained in measuring at different times the same photograph, and depend somewhat on the judgment exercised in causing the images of the position-wires to be bisected exactly by the wire of the microscope.

Photographs of the various phases of the partial eclipse, either previous to or after totality, exhibit a very curious phenomenon. The concave edge of the sun in immediate contiguity with the moon's limb, appears brighter than the other neighbouring parts of the crescent, while the convex limb of the sun bordered by the dark background of the sky, does not appear at all brighter than its proximate parts. This brightening of that part of the sun's disk which borders on the moon's limb, extends only for the space of a narrow line beyond the latter, but is remarkably conspicuous. As it cannot be accounted for by assuming the existence of a lunar atmosphere, it naturally excited a desire to trace out its cause. The Astronomer Royal, to whom I pointed out the fact, ascribed it to the effect of contrast, and I have subjected this hypothesis to the test of experiment in the following manner:—Having made some photographic prints of the sun's crescent on paper, which showed the appearance in a striking manner, I cut out about half of the crescent with sharp scissors, in such a way that the visible surface of the sun might be lifted up like a tongue, and replaced in its normal position within the background at pleasure; on smoothing the part so cut out, and causing it to occupy its original place, the bright line was apparent, but it disappeared when the crescent was lifted up, and a sheet of white paper was interposed between it and the dark ground of the photograph. These phenomena occurred when the photograph was examined with the naked eye, with the aid of spectacles, or, from a short distance, with a sharply defining telescope by Ross. Viewed in either of these ways, the brightening was found to begin immediately beyond the edge of the white paper as it was introduced more or less under the crescent.

For the purpose of illustrating this paper on the occasion of its being read before the Society, I prepared a representation of one of the photographs of partial phase, 3 feet in diameter, in which, bearing in mind the well-known fact that there is on the solar disk a gradual diminution of the intensity of the light from the centre to the periphery, I carefully reduced the brightness of the solar crescent in due gradation towards the convex boundary. In the first instance the background was not painted in, and I expected that when it was completed a brightening would immediately occur. Such, however, was not the case.

On calling Professor Stokes's attention to this failure in producing the phenomenon of brightening by artistic means, he suggested that I should renew the attempt by using a real photograph of the sun and a dark disk for the moon*. On this plan I succeeded in making eclipse-pictures artificially, which showed the brightening very distinctly. From these experiments I am inclined to believe that Mr. Airy's explanation is the true one, although it is a curious subjective fact that the parts possessing superior illumination exhibit to our perception an extremely bright line, bordering immediately on the dark limb of the moon, while the less bright parts towards the circumference present no such appearance, although they also are contrasted with the dark background.

In order to study other points connected with the photographs, I had made, on glass, some enlarged copies, in which the moon's disk was increased in some cases to 9 inches, in others to 13 inches in diameter. It was found that measurements could be made on these with considerable accuracy, by means of a graduated beam compass reading to thousandths of an inch, and I had proceeded to some extent in this way, when it occurred to me to have an instrument constructed expressly for measuring the original negatives. The study of the enlarged copies led, however, to a method of producing charts of the prominences with complete fidelity; and the plan will, I think, hereafter prove applicable to the production not only of astronomical, but also of other graphic representations derived from photographs. In order to carry out this method, a table had to be constructed with a square hole cut in it somewhat smaller than the glass positive to be worked upon; a recess surrounding the hole was made in the top of the table, just the size of the glass, and of a depth corresponding to the average thickness of the plates. Four spring clips served to hold the glass firmly in its seat. Parallel with

* For the lunar disk I employed photographic paper darkened to the same tint as the background of the solar photograph. These disks were in some cases neatly inserted in a circular hole in the solar picture, and in other cases pasted on it. In either case the surface was polished and made uniform by passing the picture through a rolling-press.

one side of the table were inserted two brass plates with long slots through which two screws worked. These screws passed through a straight edge, which could be adjusted so as to cause a right-angled drawing-triangle resting against it to assume any required position with respect to the image of the position-wires depicted on the photograph. A long glass mirror was attached to the frame of the table underneath the top, in such a way as to be adjustable to the angle best suited to reflect light through the transparent positive. In front of and above the table was placed an inclined screen formed of tissue paper, to diminish the direct light, a certain amount of which was required to show the position of the etching-point. Without the aid of this screen, the direct light would have been too powerful, and would have prevented the details of the transparent photograph from being seen by the light transmitted from below after reflexion from the mirror.

In the first place, the centre of the picture was found, and marked with a diamond point. A drawing-triangle, with one angle of 90° and two of 45°, was now placed over the photograph, with one of its sides resting against the adjustable straight edge, when its hypothenuse would coincide approximately in direction with the images of the wires. By adjusting the straight edge, the hypothenuse of the drawing-triangle was brought to exact coincidence with either of the wires, and the straight edge, against which it rested, was then (by means of the screws passing through it) clamped in position. By sliding the triangle along in contact with the straight edge, a line parallel with the wire was next set off passing through the centre, and marked slightly on the periphery of the picture by scratching with a diamond point through the collodion film. On taking in the beam-compasses a chord corresponding to 45° plus the known + or − error of the wires, a circle of right ascension, or a parallel of declination, could be made to pass through the centre, and, the points of its intersection with the lunar disk having been marked, any angles of position could be ascertained, by taking the chord between any part of a protuberance to be measured and the normal points thus set off.

If it were desired to produce an etching of any photograph, the outline of the protuberances or of the sun's disk and spots, or of the crescent of the sun, as the case might be, was traced very carefully with an etching-point through the collodion, with the aid of a lens. When this had been done, the plate was warmed by holding it before a bright clear fire, and a piece of composition, consisting of a mixture of paraffin and white wax, rubbed over it; the heat of the plate caused the waxy mixture to melt, and thus a very even, thin, and translucent etching-ground was laid on the glass. The outline was now traced a second time, in this instance through the wax, and a camel-hair pencil, wetted with liquid hydrofluoric acid, was rapidly run over the parts traced. In about a minute the acid was removed with blotting-paper, and the plate rinsed with water, and again dried with bibulous paper. When quite dry, the wax was melted by holding the plate before the fire, and wiped off with a cloth. If the etching proved satisfactory, it was again covered with an etching-ground, then centered on the circular dividing-engine, and degrees and subdivisions set off, starting from a normal point previously marked on the plate. It was then put on the straight-line engine, and a scale of minutes and seconds of arc set off from the moon or the sun's periphery, in accordance with the previously calculated value in arc of subdivisions of an inch; both sets of division were then etched in the same way as the outline.

Sometimes, according to the position in which the photograph was taken*, the etching was performed at the back of the plate, to correspond with the previous tracing through the collodion on its face. In this case the collodion picture might be allowed to remain as "a witness" (as workmen call it) of the correctness of the etching. In other cases, if the original negative had been purposely turned over, so as to present the opposite face to the camera, then the etching was made through the collodion, which had to be removed before the subsequent operations about to be described were performed.

An etched glass plate, if filled with printing ink, could be made to give a print by placing india proof-paper over it, and, after superposing a sheet of glazed paper upon this, rubbing the latter carefully with a burnisher; but it would not be advisable to attempt to take many impressions in this way. However, by the well-known processes of electrotype, copper duplicates of the glass plate can be procured, which can be printed from in the ordinary copper-plate press; and as the glass plate is only used for furnishing the matrices, and is not injured thereby, the printing-plates may be procured without practical limit as to number. In this way Plates XIII., XIV., XV., XVI. and XVII. were obtained. The original glass plate of Plate XV. was, however, made in a somewhat different manner from the others. Originally, it was a photograph of the sun; after the outline of the sun and his spots had been etched, and the normal line marked thereon, the collodion was entirely removed, to permit of the plate being superposed, accurately, first over Plate XIII., and then over Plate XIV. Previously, however, Plate XV. was coated with the transparent etching-ground, and the luminous prominences depicted on Plate XIII. traced off, care being taken to ensure the parallelism of the normal line of one plate with that of the other, and internal contact between the peripheries of the sun and moon respectively. The same thing was done with Plate XIV., the prominences visible in the two pictures being placed in coincidence. In this way the pictures of the prominences could be made to assume their proper position around the sun's picture. In order to facilitate this operation, a positive picture had been previously taken with the enlarging-camera, from both the original totality negatives laid one over the other, and combined suitably together, so as to form in one picture a correct representation of the whole of the prominences. When the two totality pictures had been traced off on Plate XV., a line was drawn to join the two positions of the moon's centre, which had been set off from Plate XIII. and Plate XIV. respectively; this line was then prolonged to show the path of the moon's centre during the period of totality; lines were also drawn to join these positions of the moon's centre, and the sun's centre, and prolonged to the periphery of the sun, to indicate the points of disappearance and reappearance of the sun's limb. When etched, this plate was

* By placing the original negative in the copying-camera with the collodion film either turned towards the lens or away from it, the picture produced was either in its natural position or reversed right for left.

angularly divided concentrically with the sun, and a scale of minutes and seconds of arc etched, starting from the sun's limb, by which means the prominences were referred to the sun's centre, and their angles of position and heights above his periphery could be read off with a fair degree of accuracy.

In the three Plates, XIII., XIV. and XV., a wrong correction was, however, applied for the errors of the wires in determining the zero of the angular divisions, namely $+23'$ for both totality pictures, instead of $+15'\cdot5$ for the first totality picture and $+16'\cdot5$ for the second; so that in taking angles of position of the prominences, the readings on Plate XIII. must be corrected by applying the number $-7'\cdot5$; those on Plate XIV. by applying the correction $-6'\cdot5$, and those on Plate XV. by applying $-7'\cdot0$.

Moreover, a small error in determining the centre in Plate XIV. also interferes with the absolute correctness of the position-angles and of the heights of the prominences above the moon's periphery. Subsequently to this being etched, I discovered the fact that the centre should have been placed about $5''$ of linear space nearer 270°, in a direction from 90° to that point, and $4''$ nearer 360°. The angular positions of some of the principal prominences, determined by measurement of the original negatives, will be hereinafter given, so that no difficulty will be experienced in correcting the position of the other prominences as read off from the Plates. The prominences in these Plates are represented in their natural (erect) position, and this is also the case with the sun-spots in Plate XV.; the position-angles are laid down from North towards East. The North point (360°) is consequently at the top, the East point (90°) is on the left hand, the South point (180°) is at the bottom, and the West point (270°) on the right hand.

In order to facilitate reference to the prominences, I have designated them on Plate XV. by capital letters, commencing with the prominence situated at right angles to the path of the moon across the sun's disk, which I call A; and I then follow on towards the east with the other capital letters, the small letters being employed, either alone, or with one or more dashes, to mark the subordinate parts.

The three principal sun-spots are marked *a*, *b*, *c* in the order of increase of their several position-angles.

In Plates XIII. and XIV. the details were drawn in on the back of the glass plate, and the collodion pictures still remain intact; Plate XV. was drawn on the face of the enlarged positive, which had been taken intentionally in a reversed position, by reversing the original negative in the copying-camera. The correction in the position of the glass negative on account of its thickness was duly made; that is to say, the totality pictures having been copied with the collodion turned towards, and the sun-picture with the collodion turned from the lens, the collodion was in this way carried from the lens a quantity equal to the thickness of the glass plate. The holder supporting the original negative was therefore moved towards the lens a similar quantity, and the relative sizes of the pictures, as a matter of course, remained undisturbed.

Plate XV. will be found useful as an index map of the prominences, and will facilitate comparisons of the results obtained by the various astronomers who observed the eclipse. Moreover, the position of the sun's axis being given on it, an idea may be formed by mere inspection, of the general distribution in heliocentric latitude of these appendages. Plate XVI. was obtained by etching a positive copy of No. 22 photograph, and Plate XVII. by etching one of No. 28 photograph. These photographs were obtained with the same setting of the copying-camera as the sun-picture, the original of Plate XV. They show the contours of the sun and moon, and furnish a means for comparing the concave outline of the luminous prominences with the profile of the moon which was in juxtaposition with it.

Reserving, for the present, an account of the measurements actually made on the engraved-glass originals of Plates XIII., XIV., and XV., I will proceed to describe a new measuring-instrument, and the measurements made by its means, not only of the totality-pictures, but also of the original negatives of the sun taken before and after the eclipse, as well as of the different phases of the eclipse. These consist of the direct measurement of the sun's radius, the direct measurement of the moon's radius (where possible), the measurement of the chord joining the cusps, the measurement of the distance between the chord and the peripheries of the sun and the moon, and also the distances of these peripheries, the determination of the angles of position of the cusps, and their angular openings, and, lastly, the position-angles of the luminous prominences. The heights of the prominences could not be determined by means of the micrometer, in consequence of an inadvertence on the part of the makers, who by mistake made it somewhat too small; here, however, the engraved-glass originals of Plates XIII., XIV., and XV. supply the numbers with sufficient accuracy, so that this oversight is of no practical moment.

Figures 1 and 2 represent the micrometer in two different positions.

Fig. 1.

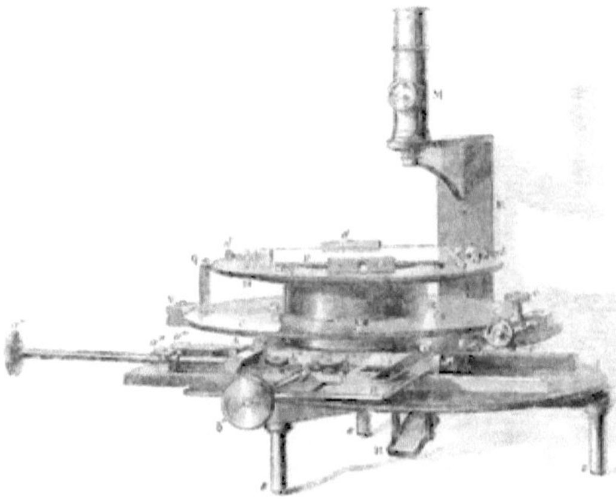

In both figures the same letter is employed to designate the same part, capital letters being used for the principal parts, and the same small letter, either alone or with one or more dashes, for the subordinate details attached to it. S is a tripod stand sup-

Fig. 2.

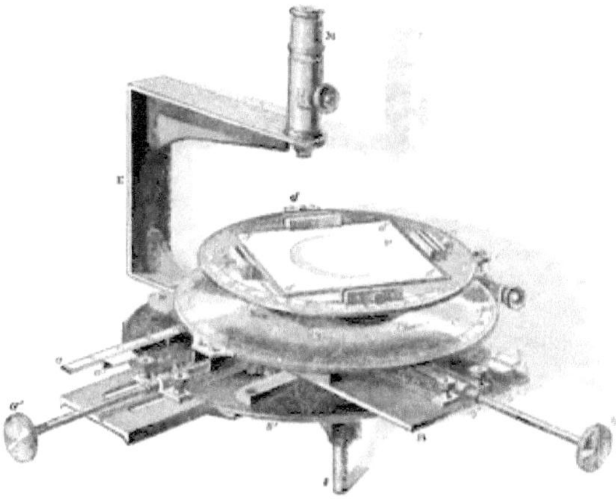

ported on the legs *s*; to this stand is firmly fixed the arm E, which supports the fixed microscope M in the centre of the instrument. This microscope can be adjusted to focus by means of the milled head shown in the engraving; and its positive eyepiece can also be adjusted by sliding it up or down, so as to bring to focus wires crossing at right angles and fixed in the centre of the field of view in such a way that their directions correspond respectively with the positions of the slides A and B, presently to be spoken of. R, fig. 1, is a plain mirror, which is adjustable so that it may reflect light through the photograph. A is a slide to which are attached all the other parts of the apparatus; it moves freely, and without vibration, between guide-bars fixed on the top plate of the tripod stand, and shown in fig. 1. The top plate of the tripod stand has a round hole in it, and the bottom plate of the slide A is perforated with a similar hole.

A steel rod a'', screwed for a certain distance at the point, works into a tapped hole in the slide A; by taking hold of the milled head a'', the slide A may be rapidly moved the whole length of the guide-bars, the rod carrying with it the clamping-piece with its two screws a''', a'''. These clamping-screws slide through two slotted holes in the top plate of the stand; when the nuts a''', a''' are screwed tight the slide is held fast, but by turning the milled head a'' the micrometer screw is also turned, and the slide can be moved for any short distance.

Attached to the slide, and moving with it, is a scale *a*, a little more than 4 inches in length, and divided into inches, tenths, and $\frac{1}{50}$ths, the inches being numbered 0 to 4. On the stand is fixed the vernier *a'* (fig. 2), which reads to $\frac{1}{1000}$th of an inch, and by estimation to $\frac{1}{2000}$th.

On the slide A are two guide-bars, between which the slide B works at right angles to A; the guide-pieces are adjustable by means of set screws *a''''* (fig. 1), to ensure the rectangularity of the slide B. Slide B is moved by means of the steel rod *b''*, screwed at one end, and carrying its clamping-bar and screws *b'''* in the same way as the slide A. It has also an attached scale *b*, and a vernier *b'*, fixed to the slide A, which reads with the scale *b* to the same quantities as the vernier *a'* of the scale *a*.

On the slide B, which is perforated, is a hollow axis, somewhat more than 4 inches in diameter, in which works the hollow axis of the divided circle C, which reads by means of its vernier V to 10'' of arc. The circle C is clamped by the clamp *c*, and can then be moved by means of the milled head *c'*, attached to a tangent screw. The vernier V and clamp *c* are fixed to the slide B.

On the hollow axis of the circle C is a second divided circle D, which reads to minutes of arc, and is divided into four quadrants; the vernier Q (fig. 1) being attached to the lower circle C. The circle D is only used for axial adjustments, to bring the position-wires depicted on the photographs to parallelism with the wires in the microscope M.

The circle D has four dogs fastened on it, through which work the screws *d*, which carry along with them four pressure plates, with two projecting wires in each, to act against the photographic plate P, and make it central with the instrument. The photograph rests on the four ivory studs *d'*.

The photograph to be measured is in all cases placed with the collodion side downwards, in order to ensure a constant distance from the microscope; if placed upwards, any variations in the thickness of the various glass plates would necessitate a change in the position of the microscope at each operation.

In the first instance it is necessary to determine that particular position of the slides A and B in which the axis of the circle C corresponds with the centre of the cross wires of the microscope, which is very accurately and rapidly accomplished in the following manner:—A glass plate of suitable size has ruled upon it with a writing-diamond two lines which intersect each other as nearly as may be at right angles: this plate is placed face downwards on the ivory supports *d'*, and the slides A and B are brought approximately to and clamped in the central position, which is at about the division 2 inches on their respective scales; by means of the screws *d*, the cross on the glass is made to coincide with the cross wires of the microscope. The circle C is now turned through half a revolution, when the cross on the glass plate will be found to have shifted. Half this deviation, according to its direction, is corrected by the screws *a''* and *b''* of the slides A and B, and half by the centering screws *d* of the circle D. After a few trials, the cross on the glass plate will not shift during the rotation of the circle C; when this is the case, the verniers of the slides A and B are observed, and the readings noted down.

The centre of the circle C (and consequently of D) was found to coincide with the cross of the microscope when A read 2·0025 inches and B 1·981 inch; and these positions were found not to alter materially for several months, during which they were from time to time tested.

The next thing to be ascertained was the rectangularity of the two slides, which was done in this way. By means of the upper circle D, one of the ruled lines on the glass plate was made to coincide with a wire of the microscope after the lower circle C had been clamped to read 360°. The slide B remaining central, slide A was unclamped and drawn out to the full extent; any deviation between the line on the glass plate and the centre of the cross in the microscope was then corrected by moving the tangent screw c', and, the slide being again pushed back to the centre, a few trials soon brought about an exact coincidence of the line on the plate and the wire of the microscope during the travelling of the slide A. The vernier V was next read off accurately, and the slide A brought to its normal position, 2·0025 inches. The circle C was then moved through exactly 90°, and clamped securely, and the slide B, having been unclamped, was pulled out to the extent of its path. If the same line on the glass plate maintained its coincidence with the cross of the microscope during the travelling of slide B, the slides were necessarily at right angles; if not, the error was corrected by moving the slide B, with respect to A, by means of the adjusting-screws a'''. If the deviation was considerable, it became necessary, after its rectification, to re-ascertain the normal central position of slides A and B. The greatest deviation from rectangularity amounted, after final adjustment, to 0° 0′ 10″, a quantity which could in no way affect the measurements.

The thickness of the wire of the microscope was ascertained to be 0·0003 inch, amounting to about 0″·15. The measurements could not generally be made nearer than 0·001 inch, equal to 0″·5, and in most cases the *centre* of the wire was brought as nearly as possible to coincidence with the point to be measured; this minute correction consequently has not been applied.

The measurement of the photographs was effected in the following manner:—In the first place, the slide B was set at its central point 1·981 inch, and the slide A pulled out so far as was judged necessary to bring the periphery of the sun (or moon, as the case might be) under the centre of the cross. The photograph to be measured was then laid on the supports, due regard being paid to the position of the wires depicted thereon, so that they might coincide with the corresponding divisions on the circle C, that is, I. with 45°, II. with 135°, III. with 225°, and IV. with 315°. The photograph was next centered by means of the four screws d, and the requisite movement of the slide A by means of its screw a'', the slide B remaining at rest. When the periphery of the sun, notwithstanding its somewhat irregular outline, maintained, during the rotation of the circle C, a general coincidence with the centre of the microscope, the vernier a' was read off; this reading, minus the reading of the central position of A, namely, 2·0025, gave the measurement of the radius of the sun. The slide A was then unclamped and traversed right across, so as to bring the limb of the sun, after passing to the other side of the centre of the instrument, into coincidence with the cross when the circle C was rotated, and the vernier a' again read off. The difference between the first and second readings gave the diameter of the sun-picture. Thus, let

	in.
The first reading of photograph No. 20 be . . .	3·907
Deduct the position of the centre on the scale *a* .	2·0025
Radius of the sun=	1·9045

	in.	
The second reading of same photograph	0·094	
And the first reading having been	3·907	
Difference between first and second reading gives . .	2)3·813	sun's diameter.
	1·9065	radius.

These measurements, in the example cited, differ from each other 0·002 inch, which, as will be hereafter seen, is equal to about 1″ of arc. In Table I., columns 7 and 8, are given two series of measurements of the sun in these ways: in column 7 the numbers were obtained by the first, in column 8 by the second method.

The next operation was the rectification of the position of the photograph by means of the wires depicted on it. I must here call attention to the circumstance, that in the various phases of the eclipse the wires cannot be traced beyond the crescent of the sun; for the time occupied in taking the photograph is so extremely short, that the background of the sky in the immediate neighbourhood of the sun does not in the slightest degree depict itself on the collodion plate. The wires are visible on the sun's crescent, because they intercept his action, and produce a blank space corresponding to their shadows, but evidently the dark body of the moon and the adjacent sky could have no such effect. In photographs of the partial phases, therefore, the length of the wires depicted corresponds with the extent of the crescent unobscured, and is shorter the nearer the picture is to the totality. A wire which is visible in some cases is afterwards covered by the moon, so that the correct apposition of the picture could not be effected, in every instance, by means of the same wire. In the totality-pictures the whole four wires are visible, from the fact of one or other of them having intercepted either the light of the prominences or that of the corona, the latter even having depicted itself sufficiently to render the shadow of the wires quite perceptible. In the other phases, either wire IV. or wire II. was used for rectifying the picture; if wire II. were used, the circle was clamped to read 134° 54′ 10″, because this had been found, by measurement, to be the position which corresponded to 315° for wire IV.

Supposing wire IV. to have been visible, and the circle clamped so as to read exactly 315° after centering, it would generally happen that the picture of wire IV. was neither under the centre of the microscope, nor parallel with the wire corresponding to the slide A, because, in the first place, the picture may not have been taken exactly in the centre of the heliograph; and in the second place, in centering, the image of the wire may not have been brought exactly parallel with a normal diameter from 315° to 135° of the circle C. If such were the case, by means of the upper circle D the image of the wire was first made approximately parallel with a wire of the microscope, and then, by means of the screw *b*″ of the slide B, was brought exactly under this wire of the microscope, so that the latter bisected longitudinally the broad image of the wire in the photograph; the slide A was now unclamped, and drawn along: if the microscope-cross continued to bisect, in the direction of its length, the wire of the photograph, the operation had been successful; if not, by again turning the upper circle D through a small arc, and moving the slide B sufficiently to cause bisection, this coincidence was finally brought about. Slide B was then screwed back to its original central position, namely 1·981 inch, and slide A to that position in which the periphery of the sun came exactly under the centre of the cross of the microscope, which in the example cited would be 3·907 inches.

The instrument was then in a position for measuring the position-angles of the cusps, which was effected by rotating the circle C so as to bring first one and then the other cusp under the microscope. Table III., columns 8 and 9, contains a series of such measurements, corrected, however, for the ascertained error of the position-wires of the heliograph, which is given in column 7, for the epoch of each photograph: the original numbers, before the correction was applied, were the means of three measurements for each cusp.

The position-angles of the cusps gave the means of finding the position-angle of the line joining the centres of the sun and moon, which is at right angles to the chord joining the cusps. For example, in photograph No. 20, the measurement of which has been quoted by way of illustration,

The position-angle of the northern cusp was found to be	13°	49′	30″*
That of the southern cusp	218	19	
Adding	360		
And dividing by 2	2)592	8	30
We obtain for the position-angle of the line joining the sun and moon's centres	296	4	15

The circle C was now fixed, so as to read the angle thus found to be that of the line joining the centres of the sun and moon, and the slides A and B were both unclamped, and so placed as to bring one of the cusps exactly to coincide with the centre of the microscope-cross. Slide A was then clamped, and slide B drawn along so as to bring the other cusp under the microscope. If it coincided with the centre of the cross, the operation was so far completed; but if not, by causing the circle C to move through a small arc, and adjusting A a little, the coincidence of both cusps was brought about. In the photograph cited, the circle had to be brought to read 296° 0′ 30″; and this, and other numbers obtained in a like manner, are the lines of centres given in Table III., column 11, corrected for the errors of the wires of the heliograph.

The coincidence of the cusps with the microscope-cross having been effected, slide B was moved so as to bring one cusp exactly under the centre of the microscope; and its position having been read off on the vernier b', the other cusp was made central, and the vernier b' again read off, the difference between the two readings giving the length

* These numbers are not corrected for the errors of the position-wires as in Table III. columns 8 and 9.

of the imaginary chord joining the cusps. Thus, in the photograph under consideration,

The first reading of vernier b' was . .	3·841
The second reading	0·124
Length of the chord = difference . . .	3·717
Length of the semichord or sine . . .	1·8585

Slide B was now brought to its normal central position, 1·981 inch, and clamped, and vernier a' of the slide A being read off, gave the position of the imaginary chord. The slide A was now moved sufficiently to bring, first, the moon's periphery and then the sun's periphery into coincidence with the centre of the microscope-cross, and the readings were recorded in each case. These readings gave the means of finding the versed sine of the moon, the versed sine of the sun, and the distance of the moon's periphery from the sun's periphery.

Thus the first reading of the vernier a' gave the position of the chord . .	2·407
The second reading the position of the moon's limb	1·154
Their difference the versed sine of the moon	1·253
Again, the first reading was	2·407
The third reading giving the position of the sun's limb	0·094
And their difference the versed sine of the sun	2·313
Position of the moon's periphery as above	1·154
Position of the sun's periphery as above	0·094
Their difference gave the distance of the sun and moon's peripheries .	1·060

Table II., columns 2 to 10 inclusive, gives a series of readings of the verniers a' and b', and the resulting determinations of the chords and the versed sines of the cusps of the sun and the moon.

By making the vernier of the circle C read 360°, plus the correction for the error of the wires of the heliograph, the position of the cusps could be read off in differences of right ascension and declination from the sun's (or the moon's) centre. Thus, in the photograph quoted, No. 20, the circle was set at 0° 9′ 30″, and the northern cusp made to coincide with the centre of the cross of the microscope by moving both the slides; the same thing was done with respect to the southern cusp. The readings were as follows:—

Northern Cusp.

Slide A	3·853	Slide B	2·429
Centre	2·0025	Centre	1·981
Difference of declination . .	+1·8505	Difference of R.A. . . .	+0·448

Southern Cusp.

Slide A	0·505	Slide B	0·803
Centre	2·0025	Centre	1·981
Difference of declination . .	−1·4975	Difference of R.A. . . .	−1·178

Although the differences of declination and right ascension of the cusps from the centres were taken out in every case, they will not be recorded in this report, because the results have been worked out by other measurements. I cite the above merely to show the applicability of the instrument to the measuring of two such coordinates as declination and right ascension.

Measurements of the moon's radius could, I found, be obtained with great accuracy by centering all the photographs which had been taken so near totality as to give a large proportion of the lunar periphery. Experience in centering for the sun had previously proved that the operation could be performed with ease, even in those cases where a large portion of the sun's disk was shut off by the moon. The totality-negatives were also measured, and I had two positive copies on albuminized glass made by superposing the negatives; and these also were measured. In the case of the totality-pictures, besides the radius, the whole diameter was measured, by causing the slide A to carry the picture from one side to the other of the centre of the reading-microscope. These measurements are given in Table I. column 9.

The readings, it will be observed, are in inches and decimals of an inch, the values of which in *arc* are based entirely on an assumed diameter for the sun, in consequence of no steps having been taken in Spain to obtain data for an absolute scale. This was not attempted, because at that time it was not contemplated that such an extensive use of the photographs would afterwards be made. It might be practicable to ascertain the value in arc of an inch on the screen of the heliograph; but this would have no retrospective application to the pictures obtained in Spain, because it would not be possible to place the object-glass, the secondary lens, and the screen in relative positions absolutely the same*. By assuming for the semidiameter of the sun its tabular value, the arbitrary measurements were translated into measures of arc; but it will be seen that any error in the tabular semidiameter of the sun necessarily affects all the numbers based upon it.

TABLE I.

In Table I. are given the following quantities:—Column 1 contains the progressive numbers of the photographs. Where a break occurs in the consecutive order, it arises from the circumstance that the omitted plate was spoiled after the picture had been taken, or that the photograph could not be taken, for the reasons already given.

Column 2 contains the times noted as the epochs of the photograph, to the nearest half-second. In observing the time, the half-second beats of the chronometer were counted, from 1 to 120, a method which, I think, is less likely to lead to error than the

* Since this paper was handed in, the Kew heliograph has been removed to Cranford, and attempts have been made to obtain an absolute value in arc for the scale of the micrometer by procuring pictures of objects situated at the distance of about a mile. Hitherto these attempts have been attended with only partial success, on account of the want of definition of the resulting pictures, consequent on the feebleness of the light and the movements of the intervening atmosphere. I have it in contemplation to continue the experiments and to erect a scale of equal parts about 80 feet long, at a known distance, with the view of ascertaining the radial distortion of the image, and the value of each increment from centre to circumference.

counting whole seconds by two steps, when using a chronometer which beats half-seconds.

Column 3 contains the computed errors of the chronometer for the several epochs, and

Column 4 the epochs of the photographs, corrected by deducting the numbers in column 3 from those of column 2.

The time-intervals in columns 5 and 6, the nature of which is sufficiently explained by their superscriptions, will be found useful in checking the numbers given in this and the subsequent Tables, they having been employed in the calculations.

Columns 7 and 8 give the results of the measurements of the sun's radius by the two methods already explained. On examining these numbers certain discrepancies will be apparent; and they may, on the whole perhaps, appear at first sight not so accordant as might have been expected. The greatest difference in column 7 is between No. 29, in which the sun's radius measured 1909·5 thousandths of an inch, and No. 16, in which it measured 1900·5 thousandths—the difference being $\frac{9}{1000}$ths of an inch, or about 4″·5. Some of this difference is due to errors in centering; for on taking the means of columns 7 and 8 for the photographs Nos. 8 and 16, the difference of radius was reduced to $\frac{7}{1000}$ths, =3″·5. Measures of the same photograph may vary in difficult cases, on account of the irregularity or faintness of the sun's border, from $\frac{1}{1000}$th to $\frac{3}{1000}$ths of an inch, but in most cases they are in complete agreement. There is, however, a real difference of photographic diameter in different pictures; for in disturbed states of the atmosphere the sun's diameter is enlarged by irradiation; and, on the other hand, when once the instantaneous apparatus has been so adjusted as to produce the best effect, any great diminution in the intensity of the light renders the picture more feeble, and the periphery of the sun consequently less distinctly defined; and it is just barely possible that the fainter portions of the limb are not depicted at all, whence would arise a diminution in the size of the picture.

The mean of the measurements given in column 7 is 1904·17, and of those in column 8, 1905·65,—the difference not being greater when converted into arc than 0″·7, while the mean of both sets of numbers is 1904·91. By assuming the radius of the sun to be 15′ 44″·8, as calculated by Mr. Hind from Leverrier's Tables, the value of $\frac{1}{1000}$th of an inch of my scale becomes 0″·4960, the logarithm of which is 9·6954654. This number has been employed in the reduction of the several measurements to their equivalents in arc.

Column 9 contains measurements of the moon, which are very accordant. The original negative of the second totality-picture presented greater difficulty in centering than the first totality-picture, in consequence of the triplication of the images of the protuberances; a positive albumen copy of it on glass was more easily centered. The greatest discordance in the measures is 4·5 thousandths, which are equivalent to 2″·2. The mean of all the measures gives 2002·25 thousandths =16′ 33″·1 as the radius of the moon, which agrees almost exactly with the number of Mr. Carrington, 16′ 33″, and that of Mr. Farley, 16′ 32″·9.

In column 10 are given the distances actually measured between the peripheries of the sun and moon in a direction at right angles to a line joining the cusps: these numbers differ in a few cases from those which were obtained for the same photographs by deducting the numbers in column 4 from those in column 3, Table II., but it has not been thought necessary to alter them in Table I. The cases in which discrepancies occur are the following:—

		inch.
No. 8.	in which the peripheral distances in Table II. differ from those in Table I. by	−0·003
No. 17.	,, ,, ,,	+0·001
No. 19.	,, ,,	−0·001
No. 22.	,, ,,	+0·003
No. 29.	,, ,, ,,	−0·001

The numbers for these particular photographs in Table II. result from a series of second measurements of several of the photographic plates, which it was found necessary to make again for Table II., in consequence of some minute errors in reading the vernier b' in the first series.

Column 11, Table I., gives the numbers of column 10 reduced to the adopted mean solar radius, namely, 1904·91 thousandths of an inch.

Column 12 gives the differences of the peripheral distances for two consecutive photographs, and, neglecting the augmentation of the moon's semidiameter, the approximate approach or retreat of the centres in the interval between their epochs. By dividing these numbers by those in column 6, were obtained the numbers in column 13, which are very nearly the rates of approach or retreat of the sun and moon's centres per minute.

Column 14 gives the approach and retreat per minute for the longer periods bracketed, and column 15 the same numbers reduced to seconds of arc. These rates of approach and retreat of the centres are affected by any errors in registering the time of the photographs, and also by all errors of measurement. The numbers do not run quite smoothly, and yet perhaps they are as good as could be expected. No account was taken of the augmentation of the moon's semidiameter, except for the middle of the eclipse; but the rates of approach or retreat for the beginning and end, even without this correction, admit of a comparison with the computed numbers, as the change of semidiameter was not great during the intervals.

	Approach of sun and moon's centres per minute at the commencement.	Relative motion of centres per minute at the middle.	Retreat of sun and moon's centres per minute at the end.
Measured . .	25″·26	27″·84	30″·05
CARRINGTON . .	24·87	27·27	29·85
FARLEY . .	25·14	27·40	29·61

TABLE I.

1. No. of photograph.	2. Observed time, Frodsham 3494.	3. Error of Frodsham 3494 at noon, 4·6 seconds fast; daily losing rate 2·1 seconds, deduced error at epoch.	4. Epochs of photographs, corrected to Greenwich mean time.	5. Interval elapsed since the epoch of No. 6.	6. Interval elapsed between two consecutive photographs.	7. Measured radius of sun, in thousandths of an inch. By measuring distance of periphery from centre.	8. By reversal and traversing slide A, to eliminate errors of centering.	9. Measured radius of the moon, in thousandths of an inch, by measuring distance of periphery from centre, &c.	10. Measured distance of sun and moon's peripheries at right angles to a line joining the cusps, in thousandths of an inch.	11. Distances reduced to the moon radius 1904·91, in thousandths of an inch.	12. Approach or retreat of peripheries in the interval, in thousandths of an inch.	13. Approach or retreat of the peripheries (or of the centres nearly) per minute, in thousandths of an inch.	14. Approach or retreat of centres per minute, calculated for the intervals bracketed, in thousandths of an inch.	15. Numbers in column 14 reduced to seconds of arc.
	h m sec./2	sec.	h m sec.	min.	min.									
6	1 47 96	+4·4	1 47 43·6	0·000		1900·5	1907·2							
7	52 71	+4·4	52 31·4	4·792	4·792	1907·5	1909·5	....	3502	3585·2				
8	56 15	+4·4	56 3·4	8·325	3·533	1900·5	1906·5	....	3403	3397·5	187·7	53·12	50·92	25·26
9	2 2 79	+4·4	2 2 35·4	14·858	6·533	1905·5	1906·0	....	3071	3072·6	321·9	49·73		
10	3 41	+4·4	3 16·4	15·542	0·684	1905·5	1900·5	....	3040	3038·2	34·4	50·29	51·71	25·65
11	7 21	+4·4	7 6·4	19·375	3·833	1902·5	1904·5	....	2840	2812·1	196·4	51·16		
12	11 31	+4·4	11 11·4	23·458	4·083	1901·5	1905·5	....	2628	2627·9	214·2	52·46		
14	20 24	+4·4	20 7·6	32·400	8·942	1904·0	1905·7	....	2176	2176·1	451·8	50·53	50·80	25·20
15	22 97	+4·4	22 44·4	35·008	2·608	1903·0	1904·7	....	2040	2041·1	135·0	51·77		
16	27 103	+4·4	27 47·4	40·058	5·050	1900·5	1905·5	....	1776	1777·8	263·3	52·11	52·19	25·89
17	33 36	+4·4	33 13·6	45·500	5·442	1900·5	1908·0	....	1489	1487·1	290·7	53·42		
19	36 43	+4·4	36 17·1	48·558	3·058	1901·5	1903·0	2001·5	1332	1333·9	153·2	50·10		
20	41 55	+4·4	41 22·1	53·658	5·100	1901·5	1906·5	2001·5	1060	1059·7	274·2	53·76		
21	46 16	+4·4	46 18·6	58·583	4·925	1904·0	1904·7	2002·5	702	702·2	267·5	54·31	53·67	26·72
22	48 60	+4·4	48 25·6	60·700	2·117	1904·5	1906·5	2001·5	680	679·8	112·4	53·09		
23 a	51 53	+4·3	51 22·2	63·643	2·943	....	....	....	523	523·0	156·8	53·28	53·28	26·43
24 b	55 86	+4·3	55 39·7	67·935	4·292	....	....	....	258	258·0	225·0	52·42		
25	First totality.	....	....	....	......		....	2001·5 d 2000·0 e	......	......	......	......	56·19	27·84 h
26	Second totality.	....	....	....	......			2001·5 f 2000·0 g	......	......	......	......		
27 c	....	....	....	....	......	....	....	....	20	204·0	......	......		
28	3 13 116	+4·3	3 13 53·7	86·108		1902·5	1903·0	2001·5	556	556·6				
29	17 32	+4·3	17 11·7	89·468	3·360	1900·5	1911·0	2003·5	740	738·0	181·4	54·97		
30	20 24	+4·3	20 7·7	92·402	2·934	1900·5	1902·0	2003·5	908	909·8	171·8	58·55	56·71	28·13
31	24 35	+4·3	24 13·2	96·493	4·091	1904·5	1905·0	2002·5	1142	1142·1	232·3	56·78		
32	26 112	+4·3	26 51·7	99·135	2·642	1902·5	1904·0	....	1289	1290·1	148·0	56·02	56·15	27·86
33	34 12	+4·3	34 1·7	106·302	7·167	1903·0	1904·7	....	1692	1692·9	402·8	56·20		
35	37 48	+4·3	37 19·7	109·602	3·300	1902·5	1904·0	....	1886	1887·6	194·7	59·00		
36	41 46	+4·3	41 18·7	113·585	3·983	1904·5	1907·0	....	2116	2115·1	227·5	57·12	57·46	28·35
37	44 1	+4·3	43 57·7	116·235	2·650	1902·5	1904·5	....	2259	2260·7	145·6	54·91		
39	51 116	+4·3	51 53·7	124·168	7·933	1903·5	1905·5	....	2720	2720·6	459·9	57·97	57·97	28·75
41	59 5	+4·2	58 58·3	131·245	7·077	1902·5	1904·5	....	3155	3137·3	416·7	58·88	58·88	29·20
42	4 2 96	+4·2	4 2 43·8	135·003	3·758	1901·5	1906·0	....	3366	3366·4	229·1	60·96	60·59	30·05
43	5 98	+4·2	5 44·6	138·020	3·017	1902·5	1906·5	....	3547	3547·8	181·4	60·13		
44	10 75	+4·2	10 33·3	142·628	4·608	1904·0	1904·0	....						
45	16 40	+4·2	16 15·8	148·397	5·769	1900·0	1907·5	....						
						1904·17	1905·65	2002·25 Moon's mean radius =log 3·3015184						

Sun's mean radius 1904·91 =log 3·2798745

$\frac{2002 \cdot 25}{1904 \cdot 91} = 1 \cdot 0511$ { ratio of lunar to solar radius.

97·3 = excess of moon's over sun's radius at the middle of eclipse.

Tabular radius of sun .... = 944″·8 = log 2·9753399
Sun's measured radius in 1000ths of an inch ... = log 3·2798745
∴ Value of $\frac{1}{1000}$th of an inch in sec. of arc 0″·4960 = log 9·6954654

Moon's radius in 1000ths of an inch .... = log 3·3015184
Value of $\frac{1}{1000}$th of an inch in sec. of arc .... = log 9·6954654
∴ Moon's radius = 993″·1 = log 2·9969838
= 16′ 33″·1

Column 11.

The numbers in this column were obtained as follows:—
Nos. in Col. 7 = r
" 8 = r′
" 10 = d
" 11 = d′

$d' = \frac{d \cdot 1904 \cdot 91 \times 2}{r + r'}$

a 23 is a solarized picture, and the contour not perfectly defined.
b 24 also is a solarized picture; and in its epoch there is reason to believe that there is some error.
c 27 is solarized; many impressions exist on the plate, in consequence of the telescope having been agitated by a gust of wind. Its epoch was not noted.
d 25. These measurements were made on the original negative.
e 25. These measurements were made on an albumen positive copy, obtained by printing from the negative by superposition (i. e. not by means of a camera).
f 26. From the original negative.
g 26. From the albumen positive.
h The augmentation of the radius of the moon, in the cases of Nos. 22 and 28 was taken account of in the computation of this quantity.

Table II.

In Table II., columns 2, 3, and 4, is given a series of measurements from which are derived the versed sines of the sun and moon, and in columns 7 and 8 other measurements from which are obtained the lengths of the chord joining the cusps given in column 9. Columns 11 and 12 the resulting semidiameters of the moon and sun respectively, calculated upon these data. On account of the change in the apparent diameter of the moon during the eclipse, the measures in column 11 are not adapted for giving a mean result of the whole series; but the case is different for the sun, and hence the average of the measures of column 12 has been taken out for comparison with the mean semidiameter obtained by direct measurement: the mean semidiameter of the sun given by Table II. is 1903·1 thousandths of an inch, which differs by only −1·91 thousandths = −0″·9 from the value given in Table I. With respect to the moon's radius, it should be borne in mind that the photographs near the commencement and the end of the eclipse are not well adapted for such calculations, as a very minute error in measuring the chord or the versed sine introduces a great error in the resulting calculated semidiameter; and any rounding off or indistinctness of the cusps, especially near the epochs of commencement and end, militates greatly against exact determinations of the moon's radius by the method employed. For these reasons, it has been necessary to omit certain numbers of column 11, in deducing the averages for the moon's semidiameter, namely, Nos. 7, 10, 14, 39, 41, 42, and 43: by bringing together into three groups the remaining calculated semidiameters of the moon, we obtain for the mean epochs of these groups the following results, as compared with those deduced from Mr. Farley's numbers for the same epochs.

		Mean epoch, $2^h\ 17^m$.	Mean epoch, $3^h\ 1^m$.	Mean epoch, $3^h\ 34^m$.
Moon's radius	De La Rue .	993″·8	993″·3	990″·9
	Farley . . .	994·1	992·9	991·9
	Difference	− 0·3	+ 0·4	− 1·0

These numbers are remarkably near the computed numbers, and render manifest that even so minute a change as the decrease in the moon's semidiameter during the eclipse is traceable in the photographs. Taking the differences of semidiameter at the first and last epochs, the augmentation of the moon's radius becomes more apparent, and not far from the true numbers; thus

		Moon's semidiameter.	
		De La Rue.	Farley.
At . .	$2^h\ 17^m$	993″·8	994″·1
At . .	$3\ \ 34$	990·9	991·9
Difference		2·9	2·2

TABLE II.—Calculations of the Radii of the Sun and Moon, from measurements of the Chords and Versed Sines.

1.	2.	3.	4.	5.	6.	7.	8.	9.	10.	11.	12.	
	Slide A.					Slide B.						
No.	Position of a line joining the cusps.	Position of moon's periphery.	Position of sun's periphery.	Versed sine of the moon =b, in 1000ths of an inch.	Versed sine of the sun =b', in 1000ths of an inch.	Position of the first cusp.	Position of the second cusp.	Length of chord, in 1000ths of an inch.	Sine, or semi-chord, =a, in 1000ths of an inch.	Calculated radius of the moon in 1000ths of an inch.	Calculated radius of the sun in 1000ths of an inch.	
7	3787	3683	91	101	3696	2613·0	1317·5	1325·5	692·7	2163·4	1907·4	A second measurement gave a similar result; the cusps are rounded off, and difficult to measure.
8	3693	3496	96	197	3597	2618·0	1114·0	1234·0	857·0	2006·3	1903·0	
9	3525	3168	94	357	3431	3117·0	835·5	2281·5	1140·7	2000·9	1905·1	
10	3504	3133	93	371	3411	3148·5	807·5	2341·0	1170·5	2032·0	1906·3	Same result on remeasurement; the northern cusp not well adapted for measurements.
11	3403	2936	96	467	3307	3267·0	694·0	2573·0	1286·5	2005·5	1903·7	
12	3294	2724	96	570	3198	3386·0	582·0	2804·0	1402·0	2009·2	1906·3	
14	3045	2271	95	774	2950	3575·0	393·0	3182·0	1591·0	2022·2	1904·0	This plate is fogged, and very difficult to measure.
15	2977	2136	96	841	2881	3615·0	353·0	3262·0	1631·0	2002·0	1902·2	
16	2925	1868	92	957	2733	3687·0	272·0	3415·0	1707·5	2001·8	1900·0	
17	2668	1563	93	1085	2573	3761·0	206·0	3558·0	1779·0	2000·9	1902·0	
19	2572	1428	97	1144	2473	3794·0	174·0	3620·0	1840·0	2003·9	1899·3	
20	2407	1154	94	1253	2313	3841·0	121·0	3717·0	1858·5	2004·8	1903·2	
21	2222	899	97	1333	2125	3870·0	93·0	3777·0	1888·5	2004·2	1904·6	
22	2133	774	91	1359	2042	3874·0	86·0	3788·0	1894·0	1999·3	1899·3	This plate difficult to measure, from the effects of atmospheric disturbance during exposure.
23	......	......	...	......	......	......	......	......	......	......	......	Solarized, and too indistinct at the cusps.
24	......	......	...	......	......	......	......	......	......	......	......	Ditto ditto.
27	......	......	...	......	......	......	......	......	......	......	......	Ditto ditto.
28	2031	655	99	1376	1932	3880·0	82·0	3798·0	1899·0	1998·5	1899·3	
29	2186	831	92	1355	2004	3880·0	87·0	3793·0	1896·5	2004·7	1905·8	
30	2308	1007	99	1301	2209	3855·0	105·0	3754·0	1877·0	2001·5	1902·0	
31	2460	1239	97	1221	2363	3829·0	133·0	3696·0	1848·0	2009·0	1904·1	
32	2547	1386	97	1161	2450	3802·0	159·0	3643·0	1821·5	2009·4	1902·1	
33	2786	1788	96	998	2690	3715·0	250·0	3465·0	1732·5	2002·8	1902·9	
35	2899	1983	97	916	2802	3658·0	308·0	3350·0	1675·0	1999·4	1901·6	
36	3022	2209	93	813	2929	3588·0	376·0	3212·0	1606·0	1992·7	1904·7	
37	3099	2355	96	714	3003	3529·0	432·0	3097·0	1548·5	1983·4	1900·7	
39	3348	2815	95	533	3253	3325·0	625·0	2690·0	1345·0	1963·5	1904·6	The southern cusp rounded, and impression faint.
41	3559	3231	96	328	3463	3057·0	901·0	2156·0	1078·0	1955·5	1899·3	An extremely faint impression, and the cusps measured with great difficulty.
42	3680	3161	95	219	3585	2869·0	1072·0	1800·0	904·0	1973·3	1906·5	A faint impression, and difficult to measure.
43	3773	3639	92	134	3681	2583·0	1285·0	1298·0	649·0	1890·3	1906·9	An extremely faint impression, and the cusps measured with great difficulty.
										Mean........	1903·1	

No.	Moon's S.D.	Sun's S. D.	No.	Moon's S.D.	Sun's S. D.	No.	Moon's S.D.	Sun's S. D.		Augmentation of moon's radius.	Farley's augmentation of moon's radius.
8	2006·3	1903·0	20	2004·8	1903·2	31	2009·0	1904·1	h m		
9	2000·9	1905·1	21	2004·2	1904·6	32	2009·4	1902·1	2 17	+0″·5	+1″·2
11	2005·5	1903·7	22	1999·3	1899·3	33	2002·8	1902·9	3 1	0·0	0·0
12	2009·2	1906·3	28	1998·5	1899·3	35	1999·4	1901·6	3 31	−2·4	−1·0
15	2002·0	1902·2	29	2004·7	1905·8	36	1992·7	1904·7			
16	2001·8	1900·0	30	2001·5	1902·0	37	1983·4	1900·7			
17	2000·9	1902·0									
19	2003·9	1899·3									
Means......	2003·8	1902·7		2002·7	1904·9		1997·8	1902·7	Between 2h 17m and 3h 34m	−2·9	
Value in arc	993″·8	943″·7		993″·3	943″·3		990″·9	943″·7			
Calculated from Farley's numbers	994·1	944·8		992·9	944·8		991·9	944·8	Ditto Farley's	−2·2	
De La Rue—Farley	−0·3	−1·1		+0·4	−1·5		−1·0	−1·1	Difference	−0·7	
Mean epoch	2h 17m			3h 1m			3h 34m				

Radius of the moon, column 11 = r.
Radius of the sun, column 12 = r'.
Versed sine of moon, column 5 = b.
Versed sine of sun, column 6 = b'.
$\frac{\text{Chord}}{2}$ of cusps, column 10 = a.

$$r = \frac{a^2}{2b} + \frac{b}{2} \qquad r' = \frac{a^2}{2b'} + \frac{b'}{2}$$

TABLE III.

In Table III., column 2, the epoch of the photographs in Greenwich mean time is again given as in Table I.; in column 3 is given a series of numbers obtained by adding to the peripheral distances of the sun and moon (as shown in Table I., column 10) the number 97·3, which expresses in thousandths of an inch the excess of the mean measured radius of the moon over the mean measured radius of the sun, namely, 2002·2 − 1904·9 = 97·3.

Column 4 contains the numbers in column 3 reduced to seconds of arc, and column 6 the same numbers corrected by the quantities in column 5, which contains the corrections necessary on account of the augmentation of the moon's semidiameter from the mean diameter at the middle of the eclipse. The numbers in column 5 are derived by interpolation from Mr. FARLEY's calculations. The corrected numbers in column 6 show the distances of the sun and moon's centres at the epochs given in column 2.

Column 7 contains the errors of the wires of the heliograph from the assumed position of 45° for wire I., for the epochs of the several photographs. These numbers have been applied to the numbers in columns 8, 9, and 10, in which, respectively, are given the corrected angles of position of the cusps, and the line joining the sun and moon's centres.

Column 12 contains the measures of half the angles between the cusps, taken from the sun's centre, the numbers being half the differences between the position-angles of each pair of cusps, which are given in columns 8 and 9. The angles in this column were employed in the computation of Table IV.

TABLE III.

1.	2.	3.	4.	5.	6.	7.	8.	9.	10.	11.	12.
No.	Epoch of photograph. Greenwich mean time.	Distance of the sun and moon's peripheries increased by 97·2, the excess of the lunar radius at the middle of the eclipse in 1000ths of an inch.	The numbers in column 3 reduced to seconds of arc.	Correction to be applied for augmentation of moon's radius interpolated from Mr. Farley's numbers.	Resulting distances of sun and moon's centres.	Error of position-wires of photoheliograph, and of their imprints on the photographs.	Position-angle of the cusps N. towards E. corrected by the quantities in column 7. Northern cusp.	Southern cusp.	Line of centres derived from the measurement of position-angles of cusps.	Line of centres ascertained by measuring the angle at right angles to an imaginary line joining the cusps.	Half the angle included between the cusps.
	h m sec.		″	″	″	′	° ′ ″	° ′ ″	° ′ ″	° ′ ″	° ′ ″
7	1 52 31·1	3682·5	1826·4	+1·4	1827·8	+ 5·4	317 17 41	276 24 41	296 51 11	a	26 26 30
8	56 3·1	3494·8	1733·4	+1·4	1734·8	+ 4·0	324 11 55	269 42 30	296 57 12	297 5 00	27 14 12
9	2 2 35·1	3169·9	1522·2	+1·3	1523·5	+ 2·0	333 46 25	259 56 55	296 51 40	296 58 40	36 54 15
10	3 16·1	3135·5	1555·2	+1·3	1556·5	+ 1·8	334 52 32	258 52 27	296 52 29	56 52	38 0 2
11	7 6·1	2939·4	1457·9	+1·3	1459·2	+ 0·7	339 18 16	254 2 48	296 40 33	42 48	42 37 45
12	11 11·1	2725·2	1351·6	+1·2	1352·8	− 0·5	344 1 45	249 16 45	296 39 15	37 39	47 22 30
14	20 7·6	2273·1	1127·6	+1·0	1128·6	− 3·2	353 22 2	239 12 57	296 32 29	28 12	56 19 32
15	22 44·1	2138·4	1060·6	+1·0	1061·6	− 4·0	355 55 00	237 31 00	296 43 00	28 00	59 12 00
16	27 47·1	1875·1	930·0	+0·9	930·9	− 5·5	0 50 45	232 49 15	296 35 00	37 00	64 15 45
17	33 13·6	1584·4	785·8	+0·7	786·5	− 7·1	6 0 21	227 6 21	296 32 21	33 6	69 27 00
19	36 17·1	1431·2	709·9	+0·7	710·6	− 8·0	8 51 31	223 42 51	296 17 11	13 30	72 34 20
20	41 23·1	1157·0	573·8	+0·5	574·3	− 9·5	13 59 00	218 28 30	296 13 45	10 00	77 45 15
21	46 18·6	889·5	441·2	+0·4	441·6	−11·0	19 50 46	212 52 21	296 21 33	26 00	83 29 12
22	48 25·6	777·1	385·4	+0·3	385·7	−11·7	21 22 37	209 34 12	295 28 24	26 52	85 54 12
23	51 22·2	620·3	307·7	+0·2	307·9	−12·6	28 24 59	200 42 42	294 33 50	b	f
24	55 39·7	395·3	196·1	+0·1	196·2	−14·0	33 53 50	190 56 47	292 25 18	c	g
27	....	117·3	58·2	−0·3	57·9	−18·0	359 41 00	260 47 00	130 44 00	d	h
28	3 13 53·7	653·9	224·3	−0·5	223·8	−19·4	30 1 19	208 45 21	119 23 21	119 24 24	80 22 2
29	17 41·7	835·3	414·3	−0·6	413·7	−20·3	34 6 18	203 18 53	118 42 35	118 41 28	84 36 17
30	20 7·7	1007·1	499·5	−0·7	498·8	−21·2	37 47 2	199 20 32	118 33 47	31 12	80 46 45
31	24 13·2	1239·4	614·7	−0·8	613·9	−22·5	42 3 30	194 17 00	118 10 15	9 50	76 6 45
32	26 51·7	1387·4	688·1	−0·9	687·2	−23·2	44 41 37	191 23 2	118 2 19	3 42	73 29 42
33	34 1·7	1790·2	887·9	−1·1	886·8	−25·4	52 14 14	183 30 49	117 52 31	117 49 24	65 38 17
35	37 19·7	1984·9	984·5	−1·2	983·3	−26·4	55 50 24	179 33 21	117 41 54	38 21	61 51 30
36	41 18·7	2212·4	1097·3	−1·3	1096·0	−27·6	60 4 36	175 16 21	117 40 28	37 36	57 35 52
37	43 57·7	2358·0	1169·5	−1·4	1168·1	−28·4	62 55 21	172 19 54	117 37 39	37 54	54 42 15
39	51 53·7	2817·9	1397·6	−1·6	1396·0	−29·8	72 30 18	162 30 3	117 30 10	32 48	44 49 52
41	58 58·3	3221·6	1604·3	−1·9	1602·4	−31·9	82 42 24	152 22 54	117 32 39	e	34 50 15
42	4 2 43·8	3463·7	1717·9	−2·0	1715·9	−33·0	88 39 00	145 33 25	117 16 12	23 30	28 17 12
43	5 44·8	3645·1	1807·9	−2·1	1805·8	−33·9	95 25 54	138 51 54	117 8 51	7 54	21 13 00

a Not measured.
b, c, d Not sufficiently distinct to ascertain the position of the line joining the cusps.
e Too faint to ascertain the position of the line joining the cusps with sufficient precision.
f, g, h The exact termination of the cusps could not be ascertained with precision.

No. 7. 296 54·2
No. 8. 296 57·2

Mean ... 296 54·2 Mean epoch h m 1 54·3

No. 9. 296 54·7
No. 10. 296 52·5

Mean ... 296 52·1 Mean epoch 2 2·9

Interval 8·6 m } Increase of position-angle per minute =0·21

No. 37. 117 37·6
No. 39. 117 30·2

Mean ... 117 33·9 Mean epoch h m 3 47·9

No. 41. 117 32·6
No. 42. 117 16·2
No. 43. 117 8·9

Mean 117 19·3 Mean epoch 4 2·5

Interval 14·6 m } Decrease of position-angle per minute 1·0

From Table III. it is possible to derive several elements of the eclipse; for example, the epochs of first and last contacts, and the duration of the eclipse.

The distance of the sun and moon's centres at the epoch of

No. 7 was .	1827″·8	No. 41 was . .	1602″·4
No. 8 was	1734·8	No. 42 was . .	1715·9
No. 9 was .	1573·5	No. 43 was . .	1805·8

whence is derived, as the mean motion per minute, in the interval between

7 and 9	=25″·263
41 and 43 . . .	=30·022

	At first contact.	Last contact.
The augmented moon's radius was	994″·5	990″·7
The sun's radius . .	944·8	944·8
	1939·3	1935·5

By deducting from these sums of the radii the corresponding distances of the centres in photographs 7, 8, and 9, and in 41, 42, and 43 respectively, and dividing the numbers so obtained by the corresponding rates of approach or retreat of the centres, the intervals are derived which have elapsed between the epochs of the first coutact and the epochs of the photographs, on the one hand; and on the other, the intervals which must have elapsed between the epochs of the photographs and the last contact of the sun and moon. By subtracting from the epochs of Nos. 7, 8, and 9, and adding to those of Nos. 41, 42, and 43 their several intervals, the following periods result:—

	First contact. h	min.	sec.		Last contact. h	min.	sec.
No. 7	1	48	6·4	No. 41 .	4	10	4·0
No. 8 .	1	47	57·4	No. 42 . .	4	10	2·7
No. 9 . . .	1	48	6·4	No. 43 .	4	10	4·0
Mean . .	1	48	3·4*	Mean . .	4	10	3·6

whence the duration of the eclipse is found to have been 2 h. 22 min. 0·2 sec.

	°	′
The position-angle of the line joining the centres at 1 h. 54·3 min. was . .	296	54·2
and by adding the decrease in the position-angle since the period of first contact, as calculated from the numbers in Table III.		1·5
we obtain for period of first contact the position-angle	296	56

which agrees very nearly with the angle calculated by Mr. Carrington and Mr. Farley, namely 296° 54′. In the same manner, the position-angle of the line joining the centres at the epochs of last contact was found to be 117° 12′

Mr. Carrington's number is . .	117	20
Mr. Farley's	117	18

* The time of first contact observed with the achromatic was 1 h. 48 m. 6·6 sec. p. 22 (354).

The beginning and end of the eclipse do not agree with the calculations; for example—

	First contact.			Last contact.			Duration.		
	h	min.	sec.	h	min.	sec.	h	min.	sec.
De La Rue . .	1	48	3·4	4	10	3·6	2	22	0·2
Carrington . . .	1	47	56	4	10	15·2	2	22	19·2
Farley	1	47	57	4	10	15	2	22	18

But it is possible that the discrepancy may partly arise in consequence of the assumption of a greater angular measure for the diameters of the sun and moon than they in reality subtend; and this view is supported by my measures of the distances of the sun and moon's centres, which, as a whole, come out greater than the computed distances. It will be seen that, as the reduction of my arbitrary measures to their equivalents in arc is dependent on the tabular value for the sun's semidiameter, if this be in excess of the truth, my distances of the centres must come out greater than the real values; and this is actually the case, as will be hereafter seen. If the diameter either of the sun or of the moon, or of both, be less than the tabular numbers, the first contact must happen later, and the last contact sooner than the computed times: the epochs of these phenomena, derived from the measure of the peripheral distances of the sun and moon given above, tend to show that some correction is necessary to the tabular diameters.

The epoch of the middle of the eclipse, the direction of motion of the moon's centre, and the nearest approach of the centres of the sun and moon may also be derived from the photographs, by means of the distances of the centres and the epochs of two photographs, one before and one after totality.

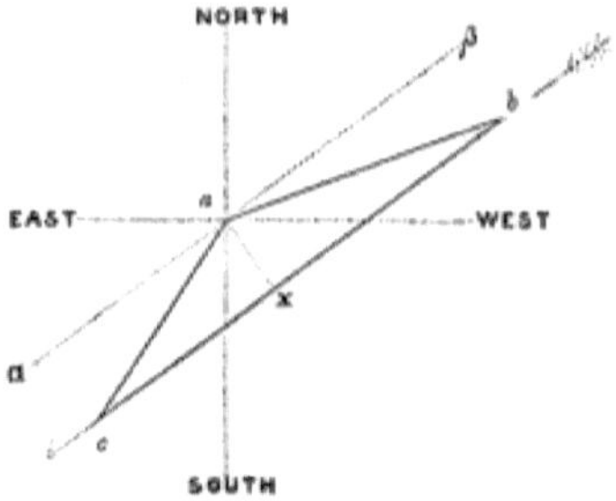

In the above diagram, let a represent the position of the sun's centre, b the position of the moon's centre previous to totality, at the epoch of No. 22 photograph, and c the position of the moon's centre in photograph No. 28, after totality; $b\,c$ will represent the motion of the moon's centre across the solar disk during the interval, and the line $\beta\,\alpha$, parallel to $b\,c$ and passing through the sun's centre, the direction of motion referred to the sun's centre; a X shows the nearest approach of the centres of the sun and moon; X the position of the centre of the moon at the middle of the eclipse.

a b, the distance of the centres at the epoch of No. 22, was	385″·7
a c, the distance of the centres at the epoch of No. 28, was	323″·8

the position-angle of the line joining the centres at the epoch of No. 22 was	205° 28′ 24″
the position-angle of the line joining the centres at the epoch of No. 28 was	119 23 21
whence the angle *c a b*	=176 5 3

	h	min.	sec.
The epoch of No. 22 was . . .	2	48	25·6
" No. 28 was . . .	3	13	53·7

and the interval, in minutes and decimals of a minute, 25·468.

	°	′	″
From these data the angle X *c a*, equal to the angle *α a c*, was found to be	2	7	43·7
And the angle X *b a*=*β a b*	1	47	13·3

The side *b* X was computed to be	385″·512
The side *c* X was computed to be	323·576
And the line *b c*, which represents the space travelled over during the interval, was consequently	709·088

	min.	sec.
The proportion of the interval of time occupied by the moon in travelling from *b* to X was computed to be	13	50·8
The proportion in travelling from X to *c*	11	37·3

The mean rate of motion per minute	0	27″·84
The nearest approach of centres, *a* X, was found to be . . .	0	12·03

	h m sec.		h m sec.
The epoch of No. 22 was . .	2 48 25·6	That of No. 28 was . .	3 13 53·7
Add time-interval *b* to X . .	0 13 50·8	Deduct time-interval X to *c*	0 11 37·3
Middle of the totality . . .	3 2 16·4		3 2 16·4

	° ′ ″		° ′ ″
The position-angle of *a b* at the epoch of 22 was . .	295 28 24	The position-angle of *a c* at the epoch of 28 was . .	119 23 21
Adding the angle *β a b* . .	1 47 13·3	Deducting the angle *α a c* .	2 7 43·7
We obtain as the direction of motion of the moon's centre during the totality from	297 15 37·3	to .	117 15 37·3

By combining Nos. 22 and 29, and Nos. 23 and 28, similar numbers were computed, which, together with the preceding results, are given in the following summary:—

	Direction of motion of moon's centre.	Nearest approach of centres.	Relative motion of centres per minute.	Middle of totality. h	m	sec.
Nos. 22 and 28 .	297° 15′ 37″·3 to 117° 15′ 37″·3	12″·03	27″·84	3	2	16·4
Nos. 22 and 29 .	297 8 53·6 to 117 8 53·6	11·27	27·78	3	2	18·4
Nos. 23 and 28 .	297 2 14·1 to 117 2 14·1	13·29	28·02	3	2	20·9
Mean .	297 8 55 117 8 55	12·20	27·88	3	2	18·6

The following are the numbers computed for the same elements by

				h	m	sec.
CARRINGTON . .	297° 17′ 0 to 117° 17′ 0	13″	27″·27	3	2	19·5
FARLEY		12·7	27·35	3	2	20·0

TABLE IV.

In Table IV. column 1, are given the computed sines of half the angles of the opening of the cusps, which are set forth in column 12, Table III.; the sun's mean measured radius, namely, 1904·91, being employed in the computations. The resulting numbers correspond to the semichords given in Table II. column 10; but in most instances the calculated is greater than the measured semichord or sine.

In column 2 are given the cosines of the same angles referred to the sun. These numbers correspond to the distance of the sun's centre from the imaginary chord joining the cusps.

In column 3 is set forth the augmented semidiameter of the moon for the epoch of each photograph, the increase or decrease from the mean measured diameter 2002·2 being derived from Mr. FARLEY's values.

Calling the sines in column $1=a$, and the augmented lunar semidiameter $=b$, the cosine of the moon for the angle represented by the same sine was derived by the formula $\sqrt{(b-a)(b+a)}$; column 4 gives these cosines referred to the lunar radius, and represents the distances of the moon's centre from the chord joining the cusps.

The distances of the sun and moon's centres could evidently be derived by taking out the sums or differences of such numbers as those in columns 2 and 4: in the cases actually under consideration, the distances of the moon and sun's centres result from the addition of these quantities; they are given in thousandths of an inch in column 5, and reduced to seconds of arc in column 6.

In column 7 are given the like quantities, derived from measurements of the distance of the peripheries, which are merely a repetition of the values given in Table III. column 6.

In column 8 are set forth the mean distances of the centres of the sun and moon, ascertained by taking the arithmetical mean between the numbers in columns 6 and 7.

In column 9 are the same distances computed by interpolation of the values calculated by Mr. FARLEY.

Column 10 gives the differences between the mean distances in column 8, and those in column 9, or De La Rue—Farley. The mean of the differences will be seen to be $+4''\cdot1$; that is, the distances of the centres of the sun and moon come out greater than the computed distances by $4''\cdot1$. This tends to show that the semidiameters of the sun and moon jointly, are less in reality by 4" than their tabular values. It is not intended to urge this as an absolute proof, but merely as supporting that view, which is further corroborated by the fact that the first contact occurred later, and the last contact sooner, than the predicted times. The distances of the sun and moon's centres, obtained by calculation from the angular opening of the cusps, will be presently employed to furnish data respecting the commencement and end of the eclipse, &c.; and it will be seen that the times thus obtained differ from those derived from the peripheral distances, and that they approach more nearly to the predicted times. The optical distortion of the sun's image would occur in the direction of a radius, and would not affect the numbers derived from measurements of the angular opening of the cusps, provided the picture were concentric with the optical axis of the instrument; while it would affect the numbers based on the measures of the distances of the sun and moon's peripheries, so that the quantity 4" is probably, from that cause, in excess of the true correction. The measurements of the angular openings of the cusps, and the measurements of the distances of the peripheries, both present peculiar difficulties. The difficulty of determining the precise termination of the cusp, especially when blunted by a lunar mountain, leads one to make the angular opening greater than it ought to be, and, consequently, the cosines and the distance of the centres less than they really are. On the other hand, the optical distortion of the image, combined with the irregularities of the peripheries of the sun and moon, tends to make the measurements of the distances of the peripheries, and consequently the distance of the centres, greater than they are in reality. These liabilities to error have, therefore, in the two cases, an opposite effect on the final results; hence a mean of the numbers obtained by the two methods will probably approach very nearly to the correction to be applied to the semidiameters of the sun and moon taken conjointly.

TABLE IV.

No.	1. Sun. Sine of ½ ∠ of opening of the cusps in 1000ths of an inch, = a.	2. Sun. Cosine of ½ ∠ of the opening of the cusps in 1000ths of an inch.	3. Mean radius of the moon, corrected for augmentation of S. D. in 1000ths of an inch, = b.	4. Moon. Cosine of ½ ∠ of the opening of cusps $\sqrt{(b-a)(b+a)}$, in 1000ths of an inch.	5. Distance of the centres of the sun and moon; col. 2 + col. 4 in 1000ths of an inch.	6. Distance of centres of sun and moon, reduced to seconds of arc.	7. Distance of centres brought from column 6, Table III	8. Distance of centres. Mean of columns 6 and 7. Seconds of arc. *	9. Distance of centres computed for epochs of photographs, by interpolating Mr. Farley's numbers.	10. Measurements minus Farley's numbers.	11. Angle of position of the line joining centres, derived from column 10, Table III. *	12. Angle of position of line of centres N. towards E. computed from Mr. Farley's numbers.	13. Measured angle of position minus Farley's angle of position of line of centres.
7	666·7	1784·9	2005·0	1899·9	3675·8	1825·1	1827·8	1825·4	1824·2	+1·2	296° 51′·2	296° 54′·0	−2′·8
8	822·1	1693·6	2005·0	1805·4	3499·0	1735·4	1734·8	1735·1	1735·4	−0·3	296 57·2	296 53·8	+3·4
9	1144·3	1533·1	2004·8	1640·3	3169·4	1572·0	1573·5	1572·7	1570·0	+2·7	296 51·7	296 52·5	−0·8
10	1172·8	1501·1	2004·8	1626·0	3127·1	1551·0	1556·5	1553·7	1552·7	+1·0	296 52·5	296 52·3	+0·2
11	1290·1	1401·1	2004·8	1534·6	2935·7	1456·0	1430·2	1457·6	1455·1	+2·5	296 40·5	296 50·9	−10·4
12	1401·6	1290·0	2004·6	1433·2	2723·2	1350·6	1352·8	1351·7	1350·6	+1·1	296 39·2	296 48·9	−9·7
14	1594·4	1042·3	2004·2	1214·4	2256·7	1119·3	1126·6	1123·9	1120·4	+3·5	296 32·5	296 42·4	−9·9
15	1636·2	975·4	2004·2	1157·4	2132·8	1057·8	1061·6	1059·7	1053·0	+6·7	296 43·0	296 40·0	+3·0
16	1715·9	827·2	2004·0	1035·2	1862·4	923·7	930·9	927·3	921·4	+5·9	296 35·0	296 33·8	+1·2
17	1783·7	668·7	2003·6	912·6	1581·3	784·3	786·5	785·4	770·7	+6·7	296 33·3	296 25·7	+7·6
19	1817·5	570·5	2003·6	843·3	1413·8	701·2	710·6	705·9	698·2	+7·7	296 17·2	296 18·5	−1·3
20	1861·6	404·0	2003·2	739·8	1143·8	567·3	574·3	570·8	563·0	+7·8	296 13·7	296 0·6	+13·1
21	1892·6	216·1	2003·0	655·8	871·9	432·4	441·6	437·0	432·1	+4·9	296 21·5	295 31·2	+10·3
22	1900·0	136·1	2002·8	633·4	769·5	341·7	385·7	383·7	375·4	+8·3	295 28·4	295 4·1	+21·3
23	....	....	....	....	....	....	307·9	307·9	296·6	....	294 33·8	291 26·5	+7·3
24	....	....	....	....	....	....	196·2	196·2	181·2	....	292 25·3	....	....
27	....	....	....	....	....	....	57·9	57·9	....	....	130 14·0	....	....
28	1904·8	21·0	2001·2	653·6	634·6	314·7	323·8	319·2	317·9	+1·3	119 23·3	120 10·7	−47·4
29	1896·5	179·1	2001·0	638·2	817·3	405·4	413·7	409·6	409·3	+0·2	118 42·6	119 7·3	−24·7
30	1880·3	305·2	2000·8	683·9	989·1	490·6	498·8	494·7	491·0	+3·7	116 33·8	118 50·3	−16·5
31	1849·2	457·2	2000·6	763·4	1220·6	605·4	613·9	609·6	605·0	+4·6	118 10·2	118 26·5	−16·3
32	1825·0	546·0	2000·4	819·1	1365·1	677·1	687·2	682·1	679·3	+2·8	118 2·3	118 17·3	−15·0
33	1735·3	785·8	2000·0	994·3	1780·1	882·9	886·8	884·8	881·6	+3·2	117 52·5	117 59·4	−6·9
35	1679·7	898·5	1999·6	1084·9	1983·4	983·7	983·3	983·5	975·7	+7·8	117 41·9	117 53·7	−11·8
36	1608·3	1020·7	1999·4	1187·8	2208·5	1095·4	1096·0	1095·7	1089·7	+6·0	117 40·5	117 48·2	−8·7
37	1544·7	1100·7	1999·2	1260·1	2368·8	1173·4	1168·1	1171·7	1165·5	+6·2	117 37·6	117 44·4	−6·8
39	1346·9	1347·0	1999·0	1477·1	2824·1	1400·7	1396·0	1398·3	1395·0	+3·3	117 30·2	117 35·4	−5·2
41	1088·2	1563·5	1998·4	1626·1	3239·6	1606·8	1602·4	1604·6	1601·6	+2·8	117 32·6	117 26·0	+4·6
42	902·7	1677·4	1998·2	1782·7	3460·1	1716·1	1715·9	1716·0	1712·3	+3·7	117 16·2	117 24·3	−8·1
43	704·8	1769·7	1998·0	1860·6	3639·3	1805·0	1805·8	1805·4	1801·3	+4·1	117 8·9	117 21·4	−12·5
										Mean + 4·1			

* Plate XVIII. exhibits a graphic representation of the path of the moon's centre, set off, in accordance with the numbers of columns 8 and 11, by means of an instrument constructed specially for such operations.

Proceeding as before, but with the distances of the centres given in Table IV., column 6, we obtain, as the rate of approach of the centres per minute between 7 and 9, 24″·945; and the rate of retreat of the centres per minute, between 41 and 43, 29″·255, whence we derive the following:

First contact.	h	m	sec.	Last contact.	h	m	sec.
No. 7 . .	1	47	51·6	No. 41 . .	4	10	12·4
No. 8 . .	1	47	52·7	No. 42 . .	4	10	13·8
No. 9 . .	1	47	51·6	No. 43 . .	4	10	12·4
Mean . .	1	47	52		4	10	12·9

Duration of the Eclipse . . . . 2h 22m 20·9sec.

These numbers agree fairly with the calculations; for example—

	First contact. h	m	sec.	Last contact. h	m	sec.	Duration. h	m	sec.
De La Rue	1	47	52	4	10	12·9	2	22	20·9
Carrington	1	47	56	4	10	15·2	2	22	19·2
Farley	1	47	57	4	10	15·0	2	22	18

	Direction of motion of the moon's centre during totality.	Nearest approach of centres.	Relative motion of the centres per minute.	Middle of eclipse. h m sec.
Nos. 22 and 29	297° 8′ 25″ to 117° 8′ 25″	11″·10	27″·34	3 2 22·6
Nos. 22 and 28	297 14 34·1 to 117 14 34·1	11·79	27·33	3 2 23·2
Mean . .	297 11 29 to 117 11 29	11·44	27·34	3 2 22·9

The following are numbers computed for the same elements by

				h m sec.
Carrington . .	297° 17′ 6″ to 117° 17′ 6″	13″·0	27″·27	3 2 19·5
Farley . . .	. . . .	12·7	27·35	3 2 20

By combining these results with the results of the peripheral measures already given, we obtain

	First contact. h	m	sec.	Last contact. h	m	sec.
By peripheries . .	1	48	3·4	4	10	3·6
By cusps	1	47	52·0	4	10	12·9
Mean . . .	1	47	57·7	4	10	8·2

Duration of the Eclipse . . . . 2h 22m 10·5sec.

	Direction of motion.	Nearest approach of centres.	Relative rate of motion of centres.	Middle of totality. h m sec.
Peripheries .	297° 8′·9 to 117° 8′·9	12″·20	27″·88	3 2 18·6
Cusps . .	297 11·5 to 117 11·5	11·44	27·34	3 2 22·9
Mean .	297 10 to 117 10	11·8	27·61	3 2 20·7

These numbers agree very closely with the theoretical numbers, the chief difference being in the epoch of the end of the eclipse, which is earlier by 7 seconds than Mr. CARRINGTON'S computation, and by 6·8 seconds than that of Mr. FARLEY.

The following are the differences:—

	De La Rue—Carrington. sec.	De La Rue—Farley. sec.
First contact	+1·7	+0·7
Last contact	−7·0	−6·8
Duration	−8·7	−7·5
Middle of eclipse	+1·2	+0·7
Nearest approach of centres	−1″·2	−0″·9
Direction of motion of the moon's centre during the totality	−7′	
Relative motion of the centres per minute	+0″·34	+0″·26
Position-angle of the line joining the centres at the first contact	+2′	+2′
Ditto at the last contact	−8′	−6′

The periods of first contact, and the middle of the eclipse, are accordant, but not so that of the end of the eclipse, the duration being less than the computed duration by 8 seconds. The half of this, or 4 seconds, would correspond to a distance moved through of 1″·9, by which quantity the radii of the moon and of the sun jointly would be smaller than the computed values. Without any desire to attach more importance to the results of the photographic measurements than they merit, I believe that I have made out satisfactorily that astronomical photography is capable of furnishing data on which great reliance can be placed, and which it would be difficult to collect in any other way. It possesses the advantage, in the case of sun-pictures, of instantaneous registration, and permits of measurements being made calmly and at leisure, and of their being repeated as often as may be considered desirable. Its employment in connexion with means of measurement will undoubtedly suggest future improvements; and although it is impossible at present to predict the destiny of astronomical photography, it appears likely that it will take a high rank among the methods of observation.

Solar Spots.

With the view of ascertaining whether any connexion exists between the luminous prominences and the faculæ, or the spots on the solar disk, photographs of the sun were obtained as soon as the heliograph could be got to work, and others would have been taken on each day previous to the eclipse if the weather had proved favourable. Several photographs were secured on the 14th, but only one on the 16th, which was not measured, having been overlooked; spots *a* and *b* were then well on the solar disk. Afterwards, until the 18th, none could be procured, but on the 19th and 20th several were obtained. With the exception of a spot *d*, which became visible on the 20th, with a position-angle 86° 26′*, a group of spots, extending on the 18th from 114° 30′ to 121° 30′†, and a double spot **X** visible on the 14th in position-angle 243° 53′ to 245° 34′‡, there were none between which and the luminous prominences any connexion could be presumed to exist. The group 114° 30′ to 121° 30′ was surrounded by many faculæ; the spots in it underwent considerable changes on the 19th and 20th; the faculæ extended evidently beyond the visible portion of the sun's surface on the 18th; for a part which was not in sight on the 18th came into view on the 19th and 20th. Just in the neighbourhood of these faculæ there was visible in the telescope during the totality, a very brilliant sheet of light. On the 18th, besides the group of spots surrounded by faculæ just mentioned, and other small spots delineated in the index map, Plate XV., there were three conspicuous spots, which I have designated by the letters *a*, *b*, *c*. Spot *c* was visible on the 14th, but the two others had not yet come round; the three spots *a*, *b*, *c* continued to be visible on the 19th and 20th.

On the whole, however, no very intimate relation was discoverable between the prominences and the sun-spots; and recent photographic researches having convinced me that the formation of spots and their changes are among the least frequent of the great disturbances always occurring in the solar photosphere, I take this opportunity of stating my opinion that future investigations will rather tend to disprove any very close connexion between them.

On the 14th, at $4^h\ 12^m\ 6^s{\cdot}6$ Greenwich mean time, the spots visible on the sun's disk were the following:—

Spot.	Position-angles.	Distance from the centre in a decimal of the radius.
c . . .	45° 37′	·3825
Cluster . α	112 21	·5548
Cluster . β . . .	113 6	·5169
Cluster . γ . . .	114 25	·4984
Cluster . δ	114 35	·5811
Cluster . ε . . .	114 51	·6075
X first nucleus . . .	243 53	·8899
X second "	245 34	·9078

The spots α, β, γ, δ, and ε, somewhat changed, were still on the disk on the 18th, but I did not notice any spot which could have been brought by rotation into proximity with the western limb of the sun, with the exception of **X**, which was at some little distance in longitude on the hemisphere turned away from the earth.

Of the group of small spots surrounded by faculæ, visible on the eastern edge of the sun on the 18th, the following were selected and measured on photograph No. 6, whose epoch is $1^h\ 47^m\ 43^s{\cdot}6$.

* See in the index map, Plate XV., the prominences E and F.

† See in the index map, Plate XV., the prominences H and G.

‡ See in the index map, Plate XV., the prominence L, which, however, was at some distance from the position of **X**.

Spot.	Position-angle.	Distance from the centre in a decimal of the radius.
ζ .	113° 33′	·9979
η .	114 40	·9953
θ .	116 31	·9643
ι . .	117 37	·9984

In consequence of the partial breaking up of the spots on the 19th and 20th, it was not easy to identify them, and the Greek letters may possibly not refer in all cases to the same spot. The following, selected from many other small spots surrounded by faculæ, were measured on a photograph taken on the 19th at 0^h 9^m 51^s Greenwich mean time:—

Spot.	Position-angle.	Distance from the centre in a decimal of the radius.
ζ .	117° 26′	·9518
θ .	119 45	·8872
ι . . .	120 19	·9607
κ	124 48	·9166

On the 20th, civil reckoning, or astronomical reckoning 19th day 23^h 46^m 45^s, the photograph K was taken, when a fresh spot *d* had made its appearance on the eastern limb; the following are the results of the measurements of this spot, and of some others in the group surrounded by faculæ. All these spots had altered greatly since the previous day.

Spot.	Position-angle.	Distance from the centre in a decimal of the radius.
d	86° 26′	·9961
ζ . . .	121 3	·8713
ι . . .	123 53	·8896
θ	126 10	·7733
κ	130 52	·8231

TABLE V.

Table V. contains the results of the measurements of the principal spots (*a*, *b*, *c*), the angles of position being corrected for the errors of the wires, and the distances given in a decimal of the radius of the sun. In the case of each spot, the particular edge measured is indicated on Plate XV. by a dotted line, and by means of Table V. my results may be reduced to those of other observers who may have measured a different part of the same spot.

Column 1 gives the number of the photograph; column 2 the date of the photograph; columns 3, 8, and 13 the distances of the spots *a*, *b*, *c* from the sun's centre in a decimal of the radius; columns 4, 9, 14 the averages of several measures; columns 5, 10, and 15 the position-angles; columns 6, 11, 16 the average position-angle for several photographs; and lastly, columns 7, 12, and 17 the mean epochs of the means of the measures.

TABLE V.

1.	2.	3.	4.	5.	6.	7.	8.	9.	10.	11.	12.	13.	14.	15.	16.	17.
					Spot *a*.					Spot *b*.					Spot *c*.	
No.	Greenwich mean time.	Spot *a*. Distance from the sun's centre in decimal of radius.		Spot *a*. Position-angle corrected for error of the wire.	Mean.	Mean epoch.	Spot *b*. Distance from the sun's centre in decimal of radius.		Spot *b*. Position-angle corrected for error of the wire.	Mean.	Mean epoch.	Spot *c*. Distance from the sun's centre in decimal of radius.		Spot *c*. Position-angle corrected for error of the wire.	Mean.	Mean epoch.
	d h m sec.			° ′	° ′	h m			° ′	° ′	h m			° ′	° ′	h m
	14 4 12 6·6	Not on disk.	...	......	...	......	......	...	......	......	......	·3625	...	45 37		
6	18 1 47 43·6	·6872		74 35·8			·6772		129 1·5			·6064		301 5·5		
7	52 31·1	·6877		75 6·5			·6798		129 16·8			·6070		300 54·3		
8	56 3·1	Defective.		74 48·3			·6761		129 16·0			·6019	·6053	301 3·0	301 0	1 54·72
9	2 2 35·1	Defective.	·6857	......	75 0	1 57·33	·6759	·6780	129 5·0	129 12	1 58·21	·6061		300 58·0		
10	3 16·1	·6868		75 8·2			·6784		129 14·3							
11	7 6·1	·6833		75 21·3			·6801		129 18·3							
12	11 14·1	·6832		75 21·0			·6764		129 23·5							
14	20 7·6	·6837	·6852	75 23·2	75 16	2 20·46	·6748	·6758	129 25·2	129 23	2 20·46					
15	22 44·1	·6867		75 3·0			·6762		129 29·0							
16	27 47·1	·6871		75 17·5			·6759		129 14·5							
17	33 13·6	·6809		75 20·6			·6752		129 41·4							
19	36 17·1	·6817		75 16·6			·6718		129 42·6							
20	41 23·1	·6825	·6821	75 0·0	75 13	2 39·31	·6730	·6737	129 31·5	129 34	2 41·13					
21	46 18·6	·6833		75 16·6			·6723		129 33·6							
22	48 25·6	......	...	......	...	......	·6730		129 22·7							
29	3 17 11·7	......	...	......	...	......	......	...	......	......	......	·6172		300 48·3		
30	20 7·7	......	...	......	...	......	......	...	......	......	......	·6170		300 22·7		
31	24 13·2	......	...	......	...	......	......	...	......	......	......	·6179	·6175	300 28·6	300 34	3 22·10
32	26 51·7	......	...	......	...	......	......	...	......	......	......	·6179		300 35·8		
33	34 1·7	......	...	......	...	......	......	...	......	......	......	·6198		300 25·4		
35	37 19·7	......	...	......	...	......	......	...	......	......	......	·6163		300 28·4		
36	41 18·7	......	...	......	...	......	......	...	......	......	......	·6187	·6189	300 30·6	300 29	3 39·16
37	43 57·7	......	...	......	...	......	......	...	......	......	......	·6207		300 31·1		
39	51 53·7	......	...	......	...	......	......	...	......	......	......	·6188		300 23·8		
41	58 58·3	Too faint	...	......	...	......	·6611		130 4·9			·6210	·6219	300 21·9	300 29	3 57·87
42	4 2 43·8	·6755		75 6·0			·6616		130 2·0			·6209		300 39·0		
43	5 44·8	·6734	·6729	74 56·0	74 53	4 8·82	·6576	·6635	130 10·9	130 10	4 6·65	·6261		300 33·9		
44	10 33·3	·6743		74 41·3			·6619		130 6·3			·6241	·6238	300 53·3	300 33	4 10·86
45	16 15·8	·6684		74 49·5			·6595		130 27·0			·6212		300 13·0		
46	19 0 9 51	·5312	...	70 48·8	...	......	·5454	...	140 18·0	......	......	·7483		296 53·0		
K	23 46 40	·3581	...	59 31·0	...	......	·4294	...	161 26·0	......	......	·8713		295 41·0		

July 18, 2ʰ 2ᵐ 35ˢ Greenwich mean time, the moon had occulted spot *c*, at a distance of ·6061 of the sun's radius from the sun's centre, and between the position-angles 300° 58′ and 302° 33′, reckoning from the respective edges of the spot.

At 3ʰ 11ᵐ 11ˢ·7 the spot *c* was partly uncovered by the moon's edge at a distance of ·6172, and between the angles 300° 48′ and 302° 45′.

At 3ʰ 58ᵐ 58ˢ·3 the moon's limb had partially passed off spot *b*, at a distance of ·6611, between the angles 128° 3′ and 130° 5′.

I am not aware that any practical use has ever been made of the occultation of a sun-spot by the moon during an eclipse; but as it is probable that during the eclipse under discussion such phenomena were recorded by a great number of observers, as on the occasion of previous eclipses, I have thought it desirable to make the measurements given in Table V., which will, I believe, afford better means than have been before available for turning such observations to account.

Photographs of the Totality.

Copies of the two totality-pictures which accompanied this paper were produced in the following way: the original negatives were placed in the focus of an enlarging-camera, and positive collodion copies on glass procured, on which the lunar disk was enlarged to 9 inches in diameter. These positive copies were then placed in the focus of the camera, and a number of negatives made, to print any impressions that might be required, in which the lunar disk was reduced to the sizes of the several engravings accompanying the paper. The photographic copies therefore are two removes from the original, and, something being lost at each operation, they do not present all the details visible in prints taken direct from the original negatives. The corona, for example, which is depicted on the original negatives, is to a great extent lost in the copies, because in bringing clearly out the details of the prominences, the corona in most cases becomes over-printed.

A few positive 9-inch copies on glass have been presented to Observatories and public Societies; but it was not possible to do this very extensively, in consequence of the extreme difficulty of copying, occasioned by the density of the original negatives. They could only be procured on days when the sun was perfectly unobscured by haze or cloud, and ultimately the injury to the second original negative prevented my continuing the work sufficiently long to obtain a supply as great as I desired. Secondary copies can, however, still be procured.

In Plate IX. are given mezzotint fac-similes of the two totality-pictures the size of the originals; although they will serve to give a general idea of the photographs, and to illustrate what has to be said respecting them, they are, after all, but imperfect substitutes for the photographs themselves. Copies of the photographs would have been inserted in this memoir, had past experience of the permanence of such pictures warranted the Council in doing so*.

* In the Author's copies, Plate IX. *a*, photographic copies from the originals are given. They are two removes from the originals.

In order that the phenomena presented by the photographs may be clearly understood, I propose to give an explanation of certain appearances in them, which might otherwise occasion some difficulty. First, with regard to No. 25 photograph (the first totality-picture) (Plate IX. fig. 1). The sensitive plate was in the heliograph and the slide which covered it removed, a minute or so before totality, a temporary screen being first held just before the object-glass, so as to stop off all light. Thus every thing was in readiness, and it was only necessary to remove the temporary screen, to expose the plate at the proper moment. The very instant of the disappearance of the sun, I gave the signal for its removal, which was immediately done, and Mr. Beckley, who was watching the chronometer, gave the signal for covering the object-glass exactly one minute after it had been uncovered. I had given instructions that no attempt was to be made to note the precise epoch of total obscuration; for each operator had too much to occupy his attention to admit of any work being done which was not absolutely essential to the photographic operations. In order to regulate the time of exposure, the precise position of the second-hand of the chronometer was noticed when the plate was first exposed, and the signal was given for replacing the screen when the second-hand had completed a revolution.

The telescope followed the motion of the sun so well that the prominences retained a perfectly fixed position on the sensitive plate; and from the results on the second plate, presently to be spoken of, it is known that they must have depicted themselves to some extent, though very faintly, in a second. On the other hand, the comparatively feeble corona would have required even a longer period to thoroughly imprint itself than the whole time allotted for the first picture. Consequently, as the moon moved from the west to the east, she kept shutting off the prominences and corona on the east, thus stopping further action; while, on the west, she permitted fresh portions of the corona to commence a new action. The luminous prominences, when once they had produced their effect, could not be obliterated, although they might be subsequently covered by the moon; for it is well known in photography that latent images remain on the plate for a long time, and become apparent on applying the developing fluid. In the case of the corona, the full effect not having been produced even at the end of the operation, it will be evident that its picture was necessarily the most intense on the eastern side, just at that part where the moon's periphery had arrived at the close of the work; while it is clear that on the western side the action would continuously follow up the moon's progress, and that therefore an impression gradually becoming fainter towards the lunar disk, would indicate the point which the moon had reached when the image of the eclipse was shut off. With this explanation, the appearance of the moon's edge beyond the prominences on the eastern side in the first totality-picture can present no difficulty. If, during the exposure of the plate, fresh prominences had become uncovered on the western side, they would have imprinted themselves; and if the plate had remained in the heliograph during the entire period of totality, the whole of the prominences would necessarily have recorded themselves on a single plate, although only a part had been visible at one time, and the plate would have shown very nearly the position of the moon at the conclusion of the operation. Unless an instantaneous picture of the phenomena of totality could be procured *, as in the case of the other phases, no photograph would show the precise state of matters at any one moment; consequently, if it be desired to know what was the condition of things at any one instant of the period during which the plate was in the heliograph, for example, at the commencement of the totality, and a minute afterwards, recourse must be had to the expedient of completing the circle of the lunar disk for the position she occupied at these two epochs respectively. The photograph itself affords the necessary data for effecting this; for it will be found that the disk of the moon, as depicted on the photograph 25, represented in Plate IX. fig. 1, is not quite a complete circle, and that the longest diameter is in a direction at right angles to prominence A (see Plate XV.).

By measuring the diameter in the direction of prominence A with a divided beam compass, and taking the half of the quantity as a radius, it was a matter of no great difficulty to find the centre of the picture for the two epochs in question, namely, near the commencement and near the end of the first minute, as indicated by the photograph, and to draw in the lunar disk from either centre. In this way two photographs, α and β, 7 inches in diameter, were corrected, and served as originals for Plates X. and XI., which show the state of the phenomena as accurately as if two instantaneous pictures had been taken.

Plate X., which is the copy of α, represents the appearance of the phenomena of totality at the commencement, and Plate XI. the copy of β nearly at the end of the first minute. A line drawn to the centres, laid down for the two epochs, was found to correspond absolutely with the direction ascertained independently to be that of the motion of the moon's centre, and measured 23″. Allowing a period of five seconds for the production of a picture sufficiently intense to show itself clearly on the plate, both at the commencement and at the end of the exposure, the period traceable would be fifty seconds, and $\frac{23'' \times 60}{50} = 27''\cdot6$ would be the motion of the moon's centre during a minute, a result not differing by more than a few tenths of a second of arc from the mean derived from the measures of the other phases of the eclipse.

In Plate IX. fig. 1, representing the untouched photograph No. 25, a portion of the prominence R is visible; but it came into view after the commencement of totality, and was therefore painted out in completing the lunar disk in the touched photograph α, represented in Plate X.; at the epoch shown in photograph β, represented in Plate XI., however, the lunar disk had revealed so much of prominence R as is seen in the original picture.

The dark lines in the original photograph, represented in Plate IX. fig. 1, situated respectively above the floating cloud C, and across the broad part of G, are the images in shadow of the position-wires. In the original negatives and positive copies taken

* The possibility of doing this during future total eclipses will be presently pointed out.

carefully from them, the continuations of the images of these wires are depicted in positions diametrically opposite. The wires are not central with the picture of the lunar disk, which was adjusted by means of the finder to be as much on the right of the plate as possible, in order that the details on the eastern limb might not be lost. It will be perceived that the light was inflected sufficiently during the taking of the picture to complete the image of the protuberance G, notwithstanding the intervening wire.

About eighty seconds were required for covering and taking out photograph 25, placing plate No. 26 in the heliograph, drawing back the slide which covered it, allowing time for the vibrations imparted to the instrument to cease, and removing the temporary cover from the telescope; so that the exposure of plate No. 26 commenced about two minutes and twenty seconds after the commencement of total obscuration, and continued until within a second of the reappearance of the sun, having been, like plate No. 25, exposed as nearly as possible one minute to the actinic influence of the prominences. The action of the prominences was, however, on account of the accidental disturbance of the heliograph, not allowed to continue on one part of the plate during the whole time; and hence their impressions are not so strongly depicted as those of the prominences on photograph No. 25, represented in Plate IX. fig. 2.

On this account the appearance of photograph No. 26 was not easy at first to comprehend; and it gave me considerable difficulty for some time to make out with precision the true nature of the result obtained. A gust of wind had arisen close upon the time this picture was being taken, and I was induced to imagine that the telescope had so been caused to vibrate; but further examination of the picture showed that this could not have been the case, for three very distinct images were imprinted of the prominence E (the boomerang), proving that after each disturbance the instrument followed the sun's movement correctly. This afforded a clue to what had actually occurred; and I found on inquiry that two of my assistants had looked at the eclipse through the finder of the heliograph, and had, as it appeared, inadvertently disturbed it in right ascension, which the wear of the worm-wheel and tangent-screw permitted them to do. Fortunately a firm radius-bar, which I had had made previous to leaving England, held the telescope so firmly in declination that it could not be readily moved in that direction; if it had been so moved, the resulting picture might have defied interpretation, and have rendered the photograph useless for exact measurements.

The clue once obtained, it was easy to make out three impressions of the wheatsheaf A (Plate XV.), three of the floating cloud D, three of each of the three points h', h'', h''' of the fallen tree H, and less easily three of the mountain-peak R because they overlapped. There were produced only two impressions of prominence Q on the western side, because at the epoch of the first action, when the point I was as yet partly visible, the moon still covered Q. The next impression was formed just when a part of Q had become visible, and when also part of the spade L had been revealed. The prominence L appears to have been sufficiently brilliant to imprint itself continuously while its image was traversing the plate. In the last impression on the plate, the prominences and corona continued their action undisturbed during the remainder of the time of exposure: this picture is the one which includes those images of the several prominences depicted furthest to the right in each case, Plate IX. fig. 2. During the exposure of the plate in its third period there was a slight irregularity in the motion of the driving-apparatus, which to a very small extent enlarged the prominences in the direction of right ascension.

In making a representation of the state of the phenomena at the end of totality, it was only necessary to paint out the two impressions of each of the several prominences belonging to the two other periods depicted on the photograph, and to correct the slight exaggeration caused by the irregularity of the driving-apparatus. In this way was produced the touched photograph γ, represented in Plate XII., which faithfully shows the state of matters about a second, or less, before the reappearance of the sun. It has been possible to make out, from photograph 26, three corrected pictures, showing the appearance of the prominences at three different epochs of that period of totality during which it was in the heliograph; but only one of the resulting pictures has been engraved, namely, that shown in Plate XII.

The irregularity of many of the protuberances on the concave side adjacent to the lunar disk is very striking, and appeared to me, while observing the eclipse with the achromatic, to be greater than could be attributable to the indentation which would be caused by any amount of irregularity on the lunar periphery. The extent of this irregularity could be readily estimated during the other phases of the eclipse with the telescope, and is also depicted clearly on the several photographs, which afford a permanent record of the moon's profile. In Plate XVI., which was produced by etching an enlarged positive copy of photograph 22 and electrotyping from it, is shown the profile of the moon's limb between the position-angles 44°·5 and 191°; in Plate XVII., produced in a like manner from photograph 28, the moon's profile is depicted between the position-angles 228° and 9°; altogether the plates exhibit 287° of the moon's outline, with which the concave edge of the luminous prominences shown in Plates XIII. and XIV. may be compared. As the moon moved onwards, the great amount of indentation of the concave side of the protuberances appeared to me to become less on the eastern side and greater on the western side. Some of the irregularity on the concave side is undoubtedly due to the periphery of the lunar disk, but all of it cannot be so accounted for. We may assume that some prominences are not in absolute contact with the sun's photosphere, but, on the contrary, are supported at a distance from it, as in the case of the floating cloud D. Notably, in that part of the prominence G between the position-angles 112° and 124° the irregularity of the concave boundary cannot be accounted for by the form of the moon's limb; on the other hand, in support of the position that in certain cases the irregularity must be due to the profile of the moon's disk, we have good evidence in the second totality-photograph, Plate IX. fig. 2, where part of the luminous prominence Q is depicted at two epochs, namely, just as it became visible, and at the end of totality. It will be seen that the part q'—q'' has the same amount of indentation at both epochs. In most cases the irregularity of the contour of the prominences appears to be much greater than that portion of the moon's limb corresponding in position-angle with it.

On comparing the results of the expedition of 1860 with those obtained in 1851, one cannot fail to be impressed with the general similarity in the aspect of the prominences at the two epochs: on both occasions were seen luminous masses of vast extent, perfectly detached from the sun, and far beyond the lunar disk; the same irregularity of outline on the convex side, running out into points; the same apparent outpouring of faint vapours, falling as it were towards the sun (as in the faint portions of prominence A); lastly, although not seen with the eye in 1860, there is recorded by the photographic retina a similar hooked projection to that seen in 1851, and named by the Astronomer Royal "the boomerang," from its approximation in form to the Australian weapon bearing that name.

Reference to Plate XV. affords better means of judging of the dimensions of the protuberances depicted on the two photographs than the photographs themselves, on account of the latter merely showing their distance from the moon's periphery, whereas in the index map, Plate XV., the distances from the sun's disk are shown. We thus see that while the two photographs give a fair representation of the height above the sun's periphery of the prominences D, E, F, G, M, N, O, P, Q, R, they do not do so in respect of the remainder. Notably, the prominence A, if it extended inwards to the sun's periphery, and was not, like the floating cloud D, supported at some distance from it, must have been of twice the height revealed to us; and again, the prominence K must, as regards the brighter part of it, have been nearly four times the height seen above the moon's limb. The brighter part of the prominence K, at the epoch of the second totality-picture, was covered by the moon; and although the fainter hooked part projected beyond the lunar disk, it was too faint to imprint itself in the second picture, on account of the disturbance of the telescope, which did not allow it to remain sufficiently long on one part of the plate. In the first picture it depicted itself clearly, and, if it had done so in the second, would have admitted, in connexion with A, of a chord being drawn which would have afforded a capital basis for measurements. The actinic power of "the boomerang" is quite remarkable; for it imprinted itself, distinctly, three times on photograph 26, although it was invisible to the eye. Not only, therefore, is photography of value in recording phenomena visible to the human eye, but it is also able to render evident bodies which emit only those rays which belong to the invisible part of the spectrum.

The question whether the luminous prominences would appear as bright or dark markings on the sun's disk admits of a probable solution by means of photography, which furnishes data as to the degree of luminosity of the prominences relatively to that of the sun's photosphere. It will be recollected that I have stated that one of the prominences (A) was visible some minutes previous to totality, and that it continued visible to other observers several minutes after the reappearance of the sun. It so happened that photograph No. 28 was taken about twenty seconds after the reappearance of the sun, when the distance between the moon and the sun's peripheries was only 9''·6. This photograph was obtained with the full aperture (3·4 inches) of the heliograph, and the plate was exposed by removing a temporary cover which had been placed before the object-glass, and replacing it as quickly as possible. The time of exposure would certainly not exceed a second, yet the image is completely solarized (bleached) from over-exposure; moreover, the wind, which rose suddenly at that period, violently shook the heliograph in the direction of right ascension, by successive impulsions against the object-end, which projected beyond the walls of the observatory; and many impressions of the solar crescent are consequently depicted on the plate, on which, however, not the slightest trace of prominence A could be made out. With the aperture of the object-glass reduced to 2 inches in diameter, and using the instantaneous apparatus, a picture of partial phase could under similar circumstances have been procured in $\frac{1}{70}$th of a second, and therefore in less than $\frac{1}{34}$th of a second with the full aperture of the telescope. Moreover, as in the second totality-picture (No. 26) the prominences were depicted three times, in consequence of two disturbances of the telescope in right ascension, during the minute the plate was exposed in the heliograph, we know that on the average twenty seconds are about sufficient to bring out the picture of the luminous prominences strongly. These triplicate images are not, however, of equal intensity, one being very faint; and therefore, assigning to this latter (what its appearance warrants) an exposure of half the time of the other two, we have twelve seconds as the time required to depict the most luminous of the prominences fairly. It results, therefore, that the light of the luminous prominences is fully $58 \times 12 = 696$ times less bright than that of the photosphere of the sun.

On August 12, 1862, I succeeded, as I have already stated in the foot-note, page 2, in obtaining an *extremely* faint impression of the moon with the Kew heliograph in three minutes. The full aperture of the object-glass was employed, and the chemicals used were in the highest degree of sensitiveness. An impression of the luminous prominences of equal intensity would, according to data furnished by the second totality-picture No. 26, have been produced in a second. It may therefore be safely estimated that the image of the luminous prominences (for equal areas) is 180 times more brilliant than that of the moon. Assuming for the ratio of the light of the sun in comparison with the light of the moon 200,000 to 1, it would follow that the image of the luminous prominences is $\frac{200,000}{180} = 1111$ times less brilliant than that of the sun; taking the mean of the two estimates it would be 900 times less brilliant than that of the sun.

Although in all probability the prominences are less bright than the dark nuclei of the solar spots, it does not follow that they would appear as very dark markings on the sun's disk, for to a great extent they may permit of the transmission of the light emitted by the photosphere; and, besides, it is by no means probable that there is any intimate connexion between the solar spots and the prominences, for the vast extent of the sun's limb which is surrounded by the prominences precludes such an idea, and leads to the conviction that they are far more generally distributed on the solar disk, and of proportions greatly exceeding any which the spots ever attain to.

Since the prominences would appear to be scattered so widely over the sun's surface, the question has arisen whether it would be possible to render these wonderful appendages apparent at other periods than those of total eclipses of the sun. For the purpose of solving this problem, Mr. JAMES NASMYTH devised an apparatus which, in part, consisted of a cylindrical box, blackened inside, having in one end an aperture of such dimensions that it exactly permitted of the passage of the sun's image when projected by a telescope, whilst the surface surrounding the aperture was sufficiently large to receive the images of all objects situated beyond the solar periphery.

The Astronomer Royal also has made experiments with the same view, using, in part, the Nasmyth apparatus; but the existence of the luminous prominences could not be detected by its means, in all probability on account of the great amount of illumination of that part of our atmosphere which is in apparent contiguity with the sun. On the occasion of Professor PIAZZI SMYTH's experimental visit to the Peak of Teneriffe he took out with him this apparatus, because it was thought that the more attenuated stratum of atmosphere at that elevation would interfere less with the success of the experiment. Only negative results were, however, obtained, and the problem remains to be solved.

Is it probable that photography may lead to a solution of the difficulty? I am inclined to think that it may possibly do so. It would, however, be quite futile to attempt to delineate the luminous prominences, when beyond the sun's periphery, by means of photography, after the experience afforded by the experiments before cited; for most unquestionably they would not produce an image so intense as that of our own atmosphere in apparent contiguity with the sun's disk and illuminated by his rays. My hope is that their forms may be depicted on the brighter solar disk itself, and their existence rendered evident by means of the stereoscope, which has already enabled me to make out the real nature of the radiating lines on the lunar surface.

During the year 1861, by means of my 13-inch equatorial reflector, I succeeded in procuring photographs of the sun's surface, on a scale of 3 feet for the sun's diameter. These colossal photographs were obtained by enlarging the focal image by means of a secondary magnifier, constructed especially to ensure a flat field and the coincidence of the visual and chemical foci. They show, in a remarkably striking manner, the mottling of the sun's photosphere, which appears to be entirely composed of an undulating mass of waves, like the surface of the sea agitated by wind.

Two pictures of the same sun-spot, taken at an interval sufficiently great to admit of the sun's rotation causing the necessary angular shift of its position, evidently possess the stereoscopic relation. By placing them in the stereoscope in such a way that the positions of the two pictures, relatively to each other, shall be reversed, that last taken being placed on the left, that first taken to the right (supposing the image to be erect), I have obtained a stereoscopic picture of a sun-spot, and some surrounding faculæ, which represented the various parts of the picture in their true relative positions in regard to altitude, and in other respects. I have ascertained in this way that the faculæ occupy the highest positions of the sun's photosphere, the spots appearing like holes in the penumbræ, which appeared lower than the brighter regions surrounding them; in one case parts of the faculæ were discovered to be sailing over a spot, apparently at some considerable height above it.

My hope of rendering evident the luminous prominences is dependent upon an extension of this experiment. I believe that, with a careful adjustment of the time of exposure of the sensitive plate, I shall succeed in obtaining the outline of the luminous prominences (the so-called red flames) as very delicate dark markings on the more brilliant mottled background of the photosphere. These delineations, except with the aid of the stereoscope, would be confounded with the other markings of the sun's surface, but they would assume their true aspect, and stand out from the rest, as soon as two suitable pictures were viewed by the aid of that instrument.

The difficulties in the way of doing this are, however, of a special kind, as will readily be seen from the following considerations: when the aperture of the instantaneous slide is at a maximum, and the rapidity of motion at a minimum, a picture of the sun will result which will be of one uniform maximum density, without the slightest trace of any marking even of a dark sun-spot. As the aperture is reduced and the velocity of the slide augmented the spots will become depicted, but no trace of the penumbræ will be seen; then we shall get the penumbræ, and subsequently traces of the faculæ and of the general mottling of the sun's disk; lastly, by still further reducing the aperture, the faculæ and the mottling will be well brought out, but especially the latter. The photographic process, it will be recollected, is one of progressive action, and even the faintest parts of the picture may, by a long exposure, produce as much intensity as the bright parts do by a shorter action; and evidently, with sufficient time, all distinctions of bright and less bright must cease to exist on the photographic plate, if all parts have produced the maximum density of effect which the plate is capable of affording. It rarely (it might be said never) happens that all parts of the picture are portrayed with the best effect; and in heliography the apparatus has to be variously adjusted, according as the spots or the mottlings of the sun's surface are required to be especially well shown.

Measurements of the Totality-Photographs.

The main object of the observations of the total eclipse of 1860 was to ascertain whether the luminous prominences are objective phenomena belonging to the sun, or whether they are merely subsidiary phenomena, produced by some action of the moon's edge on light emanating originally from the sun. If the luminous prominences are attached to the sun, it is evident that they would continually change their positions with respect to the moon's centre as the moon moved across the solar disk. For example, a prominence situated in the direction of the moon's path would maintain its position-angle in reference to the moon's centre unchanged, but it would be gradually and at length entirely covered by the moon. On the other hand, a prominence situated at right angles to the moon's path would change its position-angle in respect of the moon's centre, but would remain uncovered to almost the same extent at the end of totality as at the commencement. Moreover, prominences situated in positions intermediate between these two would change their respective position-angles, referred to the moon's centre, less, but would be covered more, in the proportion of their relative degrees of proximity to the line of the moon's path; and *vice versâ*, the change in position-angle of the several prominences would be greater in proportion as they were situated nearer to a line at right angles to the moon's path*. No stronger evidence that the prominences belong to the sun can be adduced, than that of a change in the angular position of a prominence in reference to the moon's centre, because it is not probable that *different* parts of the moon's periphery would produce precisely the same effect on light emanating from the sun. The relative changes of position may be calculated for any given locality; for which purpose it is necessary to know either its geographical position, or to determine the relative motions of the sun and moon during an eclipse by other means. In a former part of this paper I have stated that the geographical position was ascertained, also that the exact path of the moon's centre across the solar disk was made out by certain measurements of the photographs; and in Plate XVIII. I have given a graphic representation of the moon's path in reference to the sun's centre during the eclipse. Mr. CARRINGTON and Mr. FARLEY's elements of the eclipse, founded on the geographical position, have already been cited, and will be presently made use of for computing the changes of position of the several luminous prominences which should occur if they belong to the sun.

If the prominences belong to the sun, photographic images of the same protuberances taken at any one locality, at different epochs of totality, ought to coincide exactly when the photographs are superposed one over the other†; and measurements of their positions with respect to the moon's centre ought to correspond with their computed positions. Moreover, photographs taken at different places (sufficiently near in longitude) ought to agree in their details,—although, for very distant localities, it is possible that they might not do so; for it is conceivable that during long intervals a change might occur in the luminous prominences, or fresh prominences might be brought into view by the sun's rotation.

Furthermore, on consideration it will be evident that to two observers differently situated the protuberances would not have precisely the same dimensions; that is, they would appear to project more or less beyond the moon's limb. Within the zone of

* An inspection of Plate XV. will render this evident, and it will be seen that the angle γ gradually diminishes as the prominence is more distant from A.

† That this was the case was shown, on the occasion of this paper being read, by sliding the first totality-picture over the second, and projecting their images on a screen by means of an electric lamp. It was thus seen that the pictures of the several prominences correspond exactly in form and position, each to each, in the two photographs.

totality, indeed, some prominences might be visible at one station and not at another, in consequence of the parallactic shift of the moon with respect to the sun.

Photographs offer great advantages over eye observations in determining changes of position in the prominences; but nevertheless there are some difficulties even in photographic measurements.

For example, with objects which are terminated by a softened outline, there is some difficulty in determining, absolutely, where the boundary ceases to exist: this was found to be especially the case with the luminous prominences depicted on the original negative, and also with the prominences as shown in positive photographic copies taken from it by superposition on an albumenized plate. Some doubt also existed in regard to the centering of No. 26 photograph; but this was found not to affect the results so much as the uncertainty of the precise termination of the prominences, in the photographs both negative and positive.

Table VI.

Table VI. gives the results of a series of angular measures of the luminous prominences, with reference to the moon's centre, both on the original negatives and on the albumen positive copies. Columns 4, 5, 6, 7 refer to the first totality-picture; columns 8, 9, 10, and 11 to the second totality-picture. Columns 7 and 11 are the measured positions, corrected for the errors of the wires, for the epochs of the two photographs; column 12 the difference in the position-angle of certain prominences at the commencement and at the end of totality. A mere inspection of column 12 renders it evident that the nearer a prominence is situated to 27° 10′, in reference to the sun's centre (which is the case with prominence A), the greater is its angular shift in reference to the moon's centre; and the nearer a prominence is to the line of motion of the moon's centre, the less is the angular change, as, for example, the head h''' of the protuberance H.

Table VI.—Angular Position of the Prominences referred to the Moon's centre.

1.	2.	3.	4.	5.	6.	7.	8.	9.	10.	11.	12.
Letter designating protuberance on diagram.	Synonym.	The part measured.	No. 25 photograph, 1st of totality, measured with the new micrometer.				No. 26 photograph, 2nd of totality, measured with the new micrometer.				Angular motion between the epochs of the two photographs.
			Original negative.	Albumen positive (copy).	Mean.	Mean corrected for error of wires, by adding 15′·5.	Original negative.	Albumen positive (copy).	Mean.	Mean corrected for error of wires, by adding 16′·5.	
A.	Cauliflower, Wheatsheaf	a	28° 16′	27° 47′·5	27° 58′·7	28° 14′	22° 15′	22° 37′	22° 26′	22° 42′·5	5° 32′
		a′	32 1·5	31 48	31 54·7	32 10	......	26 8	26 8	26 24	
B.	Small spot	...	46 30	45 18	45 54	46 9					
C.	Detached cloud	c	58 14·5	58 2·5	58 8·5	58 24	53 25	53 43	53 34	53 50	4 34
D.	Base of Boomerang	d }	72 16·5	72 2	72 9·2	72 25	67 57	68 16	68 6·5	68 23	4 2
E.	"	e }									
F.	......	f	85 17·5	85 18	85 17·7	85 33					
H.	Part of trunk of fallen tree	h′	......	......	......	......	105 16·5	105 19	105 17·7	105 34	
		h″	......	......	......	......	106 28	106 35·5	106 31·7	106 48	
		h‴	109 30	......	109 30	109 45	108 33	108 42	108 37·5	108 55	0 52
G.	Part of long eastern protuberance	g′	134 9	134 12	134 10·5	134 26					
I.	Mitre	i′	155 1	155 3	155 2	155 17					
K.	Hook	k″	195 9	195 17	195 13	195 28					
L.	Spade	l	......	......	......	......	261 33	261 58	261 45·5	262 2	
		l′	......	......	......	......	263 24	263 18	263 21	263 37	
M.	Bead	m	......	......	......	......	275 58	275 52	275 55	276 11	
Q.	......	q	......	......	......	......	298 13	298 9	298 11	298 27	
		q′	......	......	......	......	309 35·5	309 26	309 30·7	309 47	
R.	Mountain, Ship	r	......	......	......	......	340 59·5	341 26	341 12·7	341 29	
		r′	348 26·5	348 12	348 19·2	348 35	344 26	344 51	344 38·5	344 55	3 45
	"	r″	......	......	......	......	351 3	351 22	351 12·5	351 29	

By Mr. Carrington's calculations the relative motion of the centres during totality was 92″·8 = βa, Plate XV.

The moon's radius at totality by measurements of the photographs was 993″·1 = aθ, Plate XV.

The position-angle of the moon's path βa prolonged in the direction of a was ascertained by measurements of photographs to be ...... 117° 16′

The position-angle of prominence A at the point a was ascertained by measurements of photographs to be 28 14

Whence the angle aβa, Plate XV. ...... = 88 56

With these elements the angle γ = the angular shift of prominence A during totality was found to be 5° 21′.

In a like manner, taking into account the distance of each prominence from the moon's centre, the angular shift of the other prominences was calculated, whence

Angular shift of prominence A = 5° 21′
" " " C = 4 35
" " " E = 3 46
" " " H = 0 41
" " " R = 3 56

The edge *a* of the prominence A (the cauliflower or wheatsheaf) is not far from a line at right angles to the path of the moon's centre, and is well situated, therefore, for ascertaining whether the change of position-angle accords with the demands of that hypothesis which assumes that the luminous prominences belong to the sun.

Disregarding, for the moment, the errors in the assumed places of the wires, we have the following results.

Position of the prominence A, measured on the edge *a*:—

	First totality-picture.	Second totality-picture.	Angular shift.
Original negative . .	28° 16′	22° 15′	5° 55′
Albumen positive . .	27 47·5	22 37	5 10·5
Difference . . .	− 22·5	+ 22	

Both these values of the angular shift have to be diminished by 1′, in consequence of the alteration of the position-wires of the instrument in the interval between the two epochs, and they become respectively 5° 54′ and 5° 9′·5.

Considering the difficulties experienced, from the causes before mentioned, in making measurements of position-angle, it is quite justifiable to take the mean of the above results; for, besides the uncertainty in determining the exact boundary of a prominence in the photograph, there is superadded, also, the difficulty of placing the photograph, quite correctly, in its proper angular position on the measuring-instrument, on account of the wires being very faintly imprinted on the western side of the picture in the first, and on the eastern side in the second, totality-picture.

The mean of the two measures gives 5° 32′ as the angular shift in the position of prominence A. Assuming, in accordance with theory, the motion of the moon's centre to have been 92″·8 during totality, we have 5° 21′ as the theoretical change of position-angle, which, deducted from the mean 5° 32′, gives the difference of 0° 11′.

On taking an average of the measurements of all the prominences, the difference between the measured and the computed angular shift is only half this quantity, as will presently be seen.

Table VII.

Table VII. contains the results of measurements of positive photographs 9 inches in diameter. Columns 4, 5, 6, 7, and 8 relate to No. 1 totality-picture; and columns 9, 10, 11, 12, and 13, to No. 2 totality picture. Columns 7 and 12 show, respectively, the differences in the determinations of the position-angles of the several prominences in No. 1 and No. 2 totality-pictures, on the 9-inch photograph and the original negatives and positive-albumen copies by superposition. I give preference to the measures by means of the instrument of the original negatives and the direct albumen copies, so far as regards the position-angles; but for measuring the amount of motion of the moon's centre during the totality, I am, for the reason already assigned, dependent on the 9-inch photographs. The whole difficulty, with respect to the latter, consisted in exactly ascertaining the position of the centre of the moon in the two pictures, and in measuring correctly a picture of the sun, enlarged to the same scale, with a divided beam compass, in order to obtain a value in arc for the measurements taken in inches and decimals of an inch. This accomplished, it was only necessary to measure the distance from the moon's centre or periphery of certain parts of the prominences in both totality-pictures, and to compare the results of the measures obtained on one picture with those on the other, for the purpose of ascertaining the amount of motion of the lunar disk in reference to the several prominences, and to reduce the resulting numbers to their value in arc.

The amount of motion is given in column 14 for those prominences which are visible in both photographs.

Thus, in a direction nearly at right angles to the path of the moon's centre, the apparent motion of the periphery was 1″, while at 83° from that point, or 7° from the line of motion of the lunar centre, it was 93″.

The motion of the moon's periphery, in respect of the several prominences situated at an angle θ would be $92''\cdot8^{*} . \sin(\theta-27^\circ\ 10')$, 27° 10′ being the position of a line at right angles to the motion of the lunar centre, as deduced from the photographic measures already given. For the following calculations, the angle θ was obtained from Plate XV., in which the prominences are referred to the sun's centre.

	Part measured.	Angle θ.		Measured motion.	Computed motion.	Measure —computation.
		°	′	″	″	″
Prominence A	*a*	26	9	1	1·6	−0·6
C	*c*	56	34	44	45·6	−1·6
C	*c′*	61	39	51	52·5	−1·5
E	*e*	67	4	65	59·5	+5·5
H	*h‴*	110	9	93	92·1	+0·9
R	*r*	346	14	60	60·9	−0·9
						+0·3 Mean.

From Table VI. are derived the following numbers:—

	Part measured.	Angular shift in respect of the moon's centre during totality.				
		Measured.		Calculated.		Measure — computation.
		°	′	°	′	′
Prominence A	*a*	5	32	5	21	+11·0
C	*c*	4	34	4	35	− 1·0
E	*e′*	4	2	3	46·5	+15·5
H	*h‴*	0	52	0	41·5	+10·5
R	*r′*	3	45	3	56·5	−11·5
						+ 4·9 Mean.

$=1''\cdot4$ motion of the moon's centre in excess of the computed quantity.

It would be extremely difficult to obtain more convincing proofs that the luminous prominences belong to the sun than the foregoing numbers offer; but I have still one more to bring forward.

* The relative motion of the centres during totality as calculated by Mr. Carrington.

TABLE VII.

Designating letter on Plate XV.	Name of protuberance.	Part measured.	Distances of Prominences from the Moon's Periphery and their Position-angles referred to the Moon's centre. No. 25 photograph. First totality-picture. Measures on the enlarged positive copy, the original of Plate XIII.	Numbers in column 4, corrected for error of the zero by subtracting 8'.	Mean of the measures of the original negative, and albumen print therefrom, Table VI. column 7.	Column 5 minus column 6.	Measured height of the prominences above the moon's periphery.	No. 26 photograph. Second totality-picture. Measures on the enlarged positive copy, the original of Plate XIV.	Numbers in column 9 corrected for error of zero by subtracting 7'.	Mean of the measures of the original negative and the albumen print therefrom, Table VI. column 11.	Column 10 minus column 11.	Measured height of the prominences above the moon's periphery, corrected for error of centering.	Motion of the moon in covering and uncovering prominences.	Distance and position-angles referred to the sun's centre. Both totality-pictures engraved round No. 6 photograph of sun. Plate XV. Position-angle on sun's centre by graphic process.	Height of prominences above the sun's periphery, corrected for error of centering in those cases in which the positions are derived from the copy of No. 26.
1.	2.	3.	4.	5.	6.	7.	8.	9.	10.	11.	12.	13.	14.	15.	16.
A.	Cauliflower ...	a.	28 40	28 32	28 14	+18	0 40	22 30	22 25	22 42·5	−19·5	0 39	0 1	26 3	1 18
		a'.	32 24	32 16	32 10	+ 6		26 35	26 28	26 24	+ 4			29 39	
C.	Detached cloud	c.	58 30	58 22	58 24	− 2	0 44	54 14	54 7	53 50	+17	0 60	0 44	58 31	0 54
		c'.					1 45					0 54	0 51	61 39	1 53
D.		d.	72 20	72 12	72 25	−13	0 40	68 5	67 56	68 23	−25				
E.	Boomerang ...	e.	68 50				2 31					1 26	1 5	67 4	2 40
		e'.	72 30	72 12	72 25	−13									
F.		f.	85 40	85 32	85 33	− 1								85 4	
		f².					0 38								0 39
		h'.					1 50	105 30	105 23	105 34	−11	0 12	1 38		
H.	Fallen tree ...	h''.					1 38	107 5	106 58	106 48	+10	0 3	1 35		
		h'''.	110 0	109 52	109 45	+ 7	1 33	108 55	108 48	108 53	− 5	0 00	1 33	110 9	
G.	Long protuberance ...	g.					1 3								1 2
		g'.	134 30	134 22	134 26	− 4	1 5							136 11	1 11
I.	Mitre	i.					1 6								1 28
		i'.	155 20	155 12	155 17	− 5								157 29	
K.	Hook	k.					0 44								1 38
		k'.	195 50	195 42	195 28	+14								198 19	0 54
L.	Spade	l.						262 20	262 13	262 2	+11	0 21		260 4	0 46
		l'.						264 5	263 58	263 37	+21				
M.	Bead	m.						276 40	276 33	276 11	+22	0 16		271 41	0 21
Q.		q.						298 40	298 33	298 27	+ 6	0 37		297 54	0 37
		q''.						309 40	309 33	309 47	−14	0 31		309 49	0 31
R.	Mountain	r'.	348 50	348 42	348 35	+ 7	0 00	344 45	344 38	344 55	−17	1 00	1 00	346 14	1 16
						Mean +1'·3					Mean +9'·0				

It is known that Señor AGUILAR, Professor MONTSERRAT, and Father SECCHI obtained photographs of the eclipse at Desierto de las Palmas, south of the central line, whilst my own were taken on the north of it. The latitude of my position was 42° 42′ N., that of Señor AGUILAR 40° 4′ 4″ N.; my position, measured on a meridian, is about 18′ north of the central line of the eclipse, Señor AGUILAR's about 4′ to the south of the central line.

Señor AGUILAR was so good as to send me paper copies of the four photographs taken at Desierto de las Palmas; and I have been able, in return, to send him copies of mine on glass, 9 inches in diameter. I have enlarged the copies presented to me to exactly the same dimensions (9 inches for the moon's diameter). One amongst the number, namely, the first taken, is sufficiently good, when magnified, to admit of measures being made on it with a fair amount of accuracy, although there is evidence, in the woolliness of the photograph, of the telescope not having followed the sun very well. Notwithstanding a want of precision in the details, I was able to ascertain with certainty that the distances between the point r' of R and a of A, between a of A and c of C, between c of C and g of G, between g of G and i of I, and between i of I and k'' of K, in Señor AGUILAR's photographs, correspond exactly with the same points in mine. I have already mentioned that when my first totality negative was superposed over the second the several prominences exactly coincided in their relative positions, and that the distance between any given points of two prominences on my first totality-photograph is absolutely the same in the second totality-photograph. Here we have the evidence carried still a step further; for the distances between given points in two prominences in Señor AGUILAR's photographs accord entirely with the distances between the same points on my own.

I may mention that the prominence G in Señor AGUILAR's photograph, from the commencement of the broad part on towards h of the fallen tree H, is much confused; H is not seen, because it is mixed up with G, which in consequence is as broad from h to g' as in its broadest part. With this explanation, however, there will be experienced no difficulty in comparing the photographs. The boomerang is not depicted on Señor AGUILAR's photographs. As my position was north of the central line, and the moon's centre was, as we have seen, shifted, by parallax, about 12″ below the sun's, it follows that I ought to have seen more of the prominence A, and less of I and K, than could be seen at Desierto de las Palmas; this is fully borne out by the photographs taken at the respective localities. The height of A above the moon's periphery is 40″ in my first totality-picture; in the corresponding picture taken at Desierto de las Palmas it is 32″, the difference being 8″. Prominence K in my first totality-picture, measured at k', is 44″, in Señor AGUILAR's 60″, difference 16″. The mean of the two measures $\frac{8+16}{2}=12''$, the relative parallactic displacement of the moon's disk at the two stations of Rivabellosa and Desierto de las Palmas, ascertained as nearly as the want of definition in the photograph obtained at the latter station permitted. It is probably less by about 2″ than the true displacement.

In conclusion, the two totality-pictures No. 25 and No. 26, when reduced to a suitable size and placed in the stereoscope, No. 25 on the left, and No. 26 on the right, afford a very beautiful view of the phenomena of totality, and one which could not be enjoyed by mortal eyes in looking at the real eclipse. Not only does the stereoscope render evident the fact of the moon being an object intervening between the observer and the sun, but it also shows it as a sphere. The triplication of the prominences must be corrected in No. 26; but any attempt to complete the lunar disk by painting on a positive copy, as α, β, and γ photographs, the originals of Plates X., XI., and XII., is immediately detected, and the corrected lunar disk appears perfectly flat. In placing the photographs in the stereoscope, the prominence A must be placed upwards, and at right angles to the line joining the centres of the photographs.

Appendix.

Having brought to a successful issue the photographic record of a total eclipse, it may not be out of place to point out for the guidance of others what steps I would recommend should hereafter be adopted.

In the foot-note to p. 2, I have mentioned that with my 13-inch reflector intense photographs of the moon were obtained in four seconds, and that, under precisely similar atmospheric circumstances, it required three minutes to obtain a feeble impression of the moon with the Kew heliograph, which, for the present, is mounted on an outrigger attached to the declination axis of my reflector. It will be remembered that for the totality-pictures obtained at Rivabellosa, under exceptionally favourable conditions in respect of the sun's altitude and the state of the atmosphere, the sensitive plate was exposed exactly one minute, the resulting photograph being remarkably dense, even to a fault. A picture of the moon, of greater intensity than the feeble image given by the Kew heliograph, could be obtained with my reflector in a second, so that it would produce pictures of the prominences in $\frac{1}{180}$th part of the time required by the heliograph, or in $\frac{60}{180}=\frac{1}{3}$ of a second. Making sufficient allowance for the difficulties in determining the exact ratio of actinic intensity in the foci of the two instruments, and also for a condition of the atmosphere less favourable than that under which the 'Himalaya' photographers worked, it may be safely estimated that, with a 13-inch reflector, perfect pictures of the prominences could be procured in two seconds. A 13-inch reflector would, however, be a cumbrous instrument to transport and erect at a distance from home; but a 9-inch reflector—or its equivalent, a 6-inch refractor, specially corrected for the actinic rays, is within the compass of such an expedition. These telescopes might be mounted with clockwork drivers on rigid equatorial stands, which must, however, be so designed as to admit of an adjustment of the polar axis to suit various latitudes. For each instrument an observatory should be constructed to take to pieces. Each observatory would require not less than four plate-holders, and about six baths to contain nitrate of silver, which must have been carefully fused*. The collodion employed should be iodized a month before use with the cadmium iodizer, and, before starting, carefully decanted into clean vessels, tied over with bladder to prevent evaporation.

I would recommend that three instruments, having a focal length of about 10 feet, should be prepared and kept, with the portable observatory, in readiness to be placed at the disposal of any such expedition as that organized by the Astronomer Royal. These instruments would give pictures of the luminous prominences in four seconds at the outside; and under favourable atmospheric conditions, in less than a second.

Observers intending to take charge of an instrument should practise with it in taking lunar photographs, previous to starting, so as to familiarize themselves completely with it. Not fewer than four persons should accompany each telescope, and two of them ought to be accomplished photographers.

Pictures of the partial phases would be best obtained with such an instrument as the Kew heliograph.

Lastly, I have deemed it to be desirable that positive copies of the eclipse-pictures should be placed in some Institution readily accessible to the public, and I have therefore presented to the South Kensington Museum a series of enlarged positive copies on glass, 9 inches in diameter, which are at present exhibited in the International Exhibition, Class XIII.

* "Report on Celestial Photography," by the author, in the Report of the British Association for 1859.

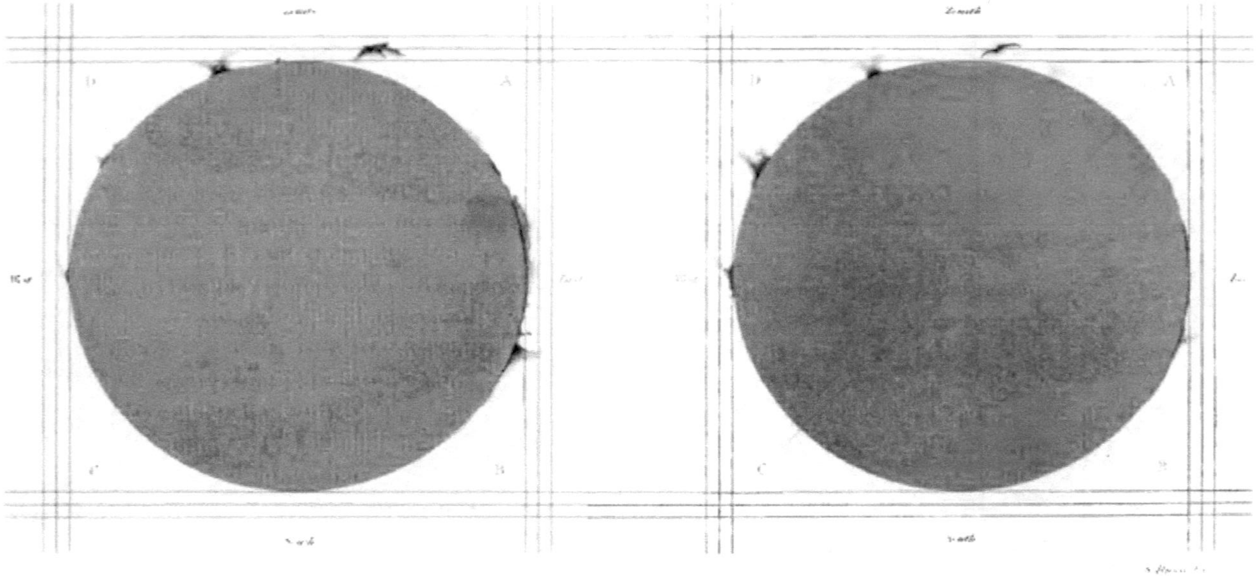

Facsimile of hand drawing

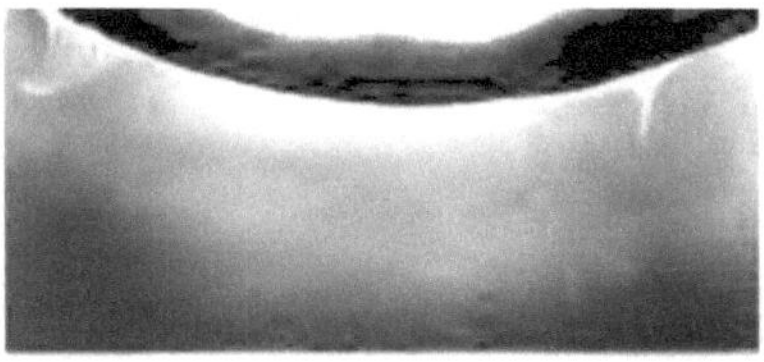

APPEARANCE OF PHENOMENA
IMMEDIATELY AFTER THE BEGINNING OF TOTALITY

APPEARANCE OF PHENOMENA
IMMEDIATELY PREVIOUS TO THE END OF TOTALITY

FAC SIMILE OF N. 25 PHOTOGRAPH (FIRST TOTALITY)

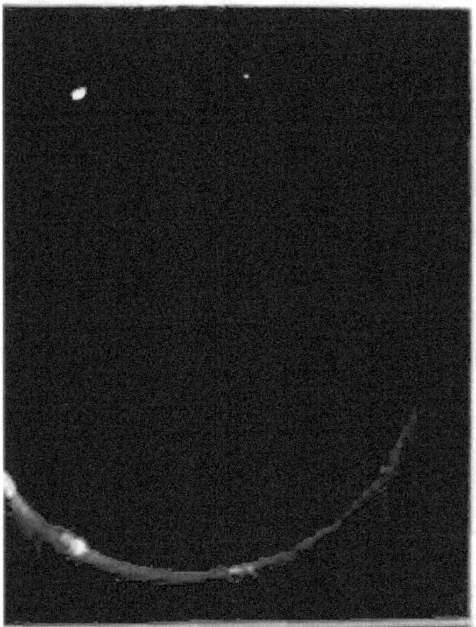

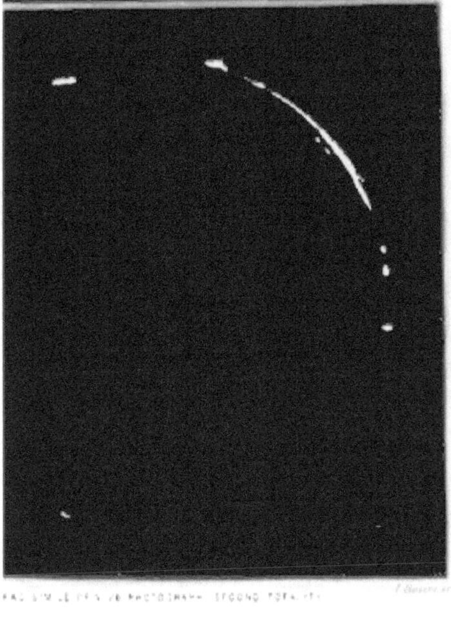

FAC SIMILE OF N. 26 PHOTOGRAPH (SECOND TOTALITY)

FIRST TOTALITY PHOTOGRAPH. [Phil. Trans. MDCCCLXII.

COPY OF A TOUCHED PHOTOGRAPH

Showing the phenomena of totality immediately after total obscuration

COPY OF A TOUCHED PHOTOGRAPH

Showing the phenomena of totality one minute after total obscuration

COPY OF A TOUCHED PHOTOGRAPH

Showing the phenomena of totality immediately previous to the reappearance of the Sun

ELECTRO DUPLICATE OF A PHOTOGRAPH ETCHED ON GLASS

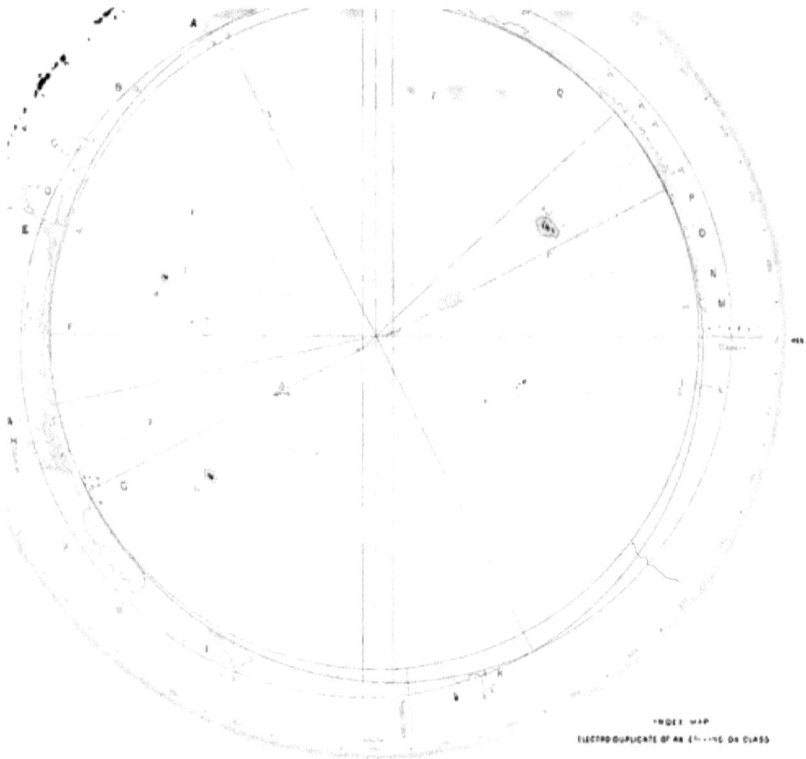

INDEX MAP

ELECTRO-DUPLICATE OF AN ENGRAVING ON GLASS

PARTIAL PHASE

[illegible]

ELECTRO DUPLICATE OF A PHOTOGRAPH ETCHED ON GLASS

PATH OF THE MOON'S CENTRE DURING THE ECLIPSE OF JULY 18th 1860, AND POSITIONS OF THE PRINCIPAL SUN-SPOTS

[illegible]

SCALE [illegible]

www.ingramcontent.com/pod-product-compliance
Lightning Source LLC
Chambersburg PA
CBHW020136170426
43199CB00010B/764